The
Track
Day Manual

Acknowledgements

First of all, a very big thank you to Sam Dorrington for the diagrams, and to John Colley for the bulk of the photographs; both have done a great job.

A similarly-sized thank you should go to those track day operators who gave advice and allowed us out on their days to take photos: Calum Lockie and Melindi Scott at Gold Track; Alex Baker at Motor Sport Vision; and Colin Jebson and Bob Willatt at Javelin Track Days. Thanks should also go to Caterham and Honda for supplying us with cars – a Civic Type R and a Roadsport – both of which performed superbly out on track.

I would also like to express my gratitude to Mark Hales for his advice on some of the elements of track technique. Thanks also to Mark Gillam, at track day prep' wizards Abbey Motorsport, for reading over the tech' chapters, and Justin Everitt at Everitt Boles (Moris.co.uk), for checking over the chapter on insurance.

Others worthy of a big thank you for helping with advice, pictures, or getting things done, include, in no particular order: Luke Brackenbury, Bob Rice, Phil Royle, Dave Woodall, Jakob Ebrey, Steve Clark, Phil Wright, Graham Clarke at RMA, Kenny P, Dave Wigmore, Volkswagen Racing, Demon Tweeks, Simon Slade at RPM, Lydden Hill, Steve Kirk, Andy Noble, Graham Cox (CG-Lock), Steve Bennett, Paul Cowland at Pro-motiv, Rick Cuthbert at Santa Pod, John Paulding, Simon McBeath, Robert Evans and Jassy.

Oh, and a special mention to all those Pug 306 drivers at Brands – for helping us solve the problem of a shortage of pictures of cars spinning!

Mike Breslin

Published in June 2008

A catalogue record for this book is available from the British Library

ISBN 978 1 84425 482 8

Library of Congress catalog card no. 2007943110

Haynes Publishing, Sparkford, Yeovil,
Somerset BA22 7JJ, UK
Tel: +44 (0) 1963 442030
Fax: +44 (0) 1963 440001
E-mail: sales@haynes.co.uk
Website: www.haynes.co.uk

Haynes North America, Inc.,
861 Lawrence Drive, Newbury Park,
California 91320, USA

Printed and bound by J.H.Haynes & Co Ltd,
Sparkford, Yeovil, Somerset BA22 7JJ, UK

Design Camway Creative Limited

Photography All photographs copyright John Colley unless otherwise indicated

Diagrams Sam Dorrington

The
Track
Day Manual

The complete guide to taking
your car on the race track

Mike Breslin

Introduction

'So you're writing a book, eh – what's it about?'

'Track days.'

'Oh, car racing.'

'No, not racing – track days.'

'Same thing; bunch of rich kids thrashing around a race track in expensive cars...'

'You're wrong there, mate; you don't need to be rich, there's no racing allowed, and you certainly don't have to have an expensive car – in fact, you could do a track day in any car, even yours.'

'What, my old thing? Now you're pulling my leg... Same again?'

There were quite a few conversations like the one to the left, not all of them in pubs, which made me realise just how little most people know about track days. That surprised me, to be honest, but then I've been writing about them for years now, and when you're closely involved you can forget there are people who don't even know what a track day is. But more to the point, you can forget that there are people who don't know what they are missing.

Most keen drivers think that it's impossible to have a little fun in their cars these days, driving on the road has become a chore, and a stressful chore at that, but thanks to track days it doesn't have to be like this. You see, the track offers an escape. It offers freedom from speed cameras, freedom from traffic congestion, freedom from road-clogging caravans, freedom from narrow-minded drivers who think it's a crime to overtake, freedom from… Well, you get the picture, the list of 21st-century petrolhead frustrations could fill this book on its own, but the point is you can get away from it all by going on a track day.

Little wonder it's such a fast growing leisure pursuit, then. But note here, that's 'leisure' not 'sport', for reasons we will go into at the very start of this book. But, if it helps, think of it as like going skiing rather than going downhill racing, and you'll have some idea of what it's all about. Just realise from the very start, a track day is not a race.

But a track day does give you the opportunity to drive your car to its max, and driving a car at the limit on the track is an incredible sensation. There's nothing quite like the feeling of balancing your car on the edge of adhesion in a fast corner. There's also the adrenalin fix you get from driving on the edge and then there's the indescribable feeling of satisfaction when you get a corner *just* right. On top of all that, there's the fact that the more you drive on track then the better driver you will become, so that if things do go wrong on the road, well then there's a fair chance you will know what to do.

These days the track is the only place for you to go if you really want to find out what your car is capable of too. Chances are, if you've a modern performance car, you've only scratched the surface of its potential, and it deserves more than the odd quick blast between knots of traffic, doesn't it?

But that's all very well and good, you're thinking, but what if my car just isn't quick enough? No worries, you don't need a supercar for track days, anything will do, as long as it's not falling apart, of course, and it doesn't need to be modified in any way either. Something sporty would be good, but it's not essential, especially when you're learning about track driving. Anyway, when it comes to sheer fun, cheap and cheerful is often the best way to go. I've been lucky enough to get out in some very quick track cars in my time as a motoring journalist, but I can honestly say that very few have matched my almost-standard £2,000 Mazda MX5 'nail' for track day *fun*.

It's not just about fun, though; let's not forget the thrills. Driving on a track is one of the few real adventures left in this antiseptic age – yet while it will give you a buzz like no other, it's also remarkably safe. I'm not saying it's completely safe – that would be silly – and if there wasn't an element of risk, driving close to the limit would probably lose its appeal, wouldn't it? That said, as long as you drive within *your* limits, and build up your speed as you build up your experience, then you'll be OK.

One thing is for sure, driving fast on the track is definitely safer than driving fast on the road, and if things do go wrong there are usually gravel traps or run-off areas to stop you before you hit something solid. It's probably safer than racing, too, mainly because there's usually a strict rule about no overtaking in corners and in the braking area into corners – the places where there's more likelihood of mishaps.

And if you're worried that *you* won't be fast enough to drive on a track day, forget it. You can drive as fast – or as slow – as you please, and

LEFT Yellow peril: just one of the reasons you *need* to go on a track day.

BELOW A race track gives you the space, and the freedom, to find out what your car is truly capable of.

anyway it's much better to build up your speed progressively. As long as you keep your eyes on your mirrors and let faster drivers through on the straights, you'll do just fine.

Some people, many of them in the world of motor racing, don't understand why people who do track days just don't go racing instead. Well, beyond the obvious – racing can be very expensive – there's also the fact that in many ways track days actually have more to offer. Certainly more in the way of track time, for where you'll be lucky to get a half-hour of qualifying and a ten-lap race at a race day, you'll get a whole day at a track day.

There's more to it than just the extra track time, though, and the reason why many accomplished

track drivers are happy enough to stick with their pursuit and not move up into racing is that track days are far more relaxed occasions than race meetings. You can go at your own speed, maybe hold back if you're a little nervous of a certain corner, and if you do have a bit of a moment, then you can enjoy it for the adventure it is – not curse yourself silently for losing a tenth of a second to the bloke in front. On top of all that, track days also tend to be relatively hassle free – just turn up, sign on, and drive.

Don't get me wrong, I love racing, and competed myself for some years. It's just that track days are *different*, and in my experience those in the racing world who knock them tend to be the ones who haven't actually been on a track

BELOW There's far more speed to be found in improving your driving technique to begin with.

day. So, yet more people who do not know what they are missing, then…

This book

One of the great things about track days is the simplicity. All you need is a drivers' licence, a helmet and a car and you're away. Nothing could be simpler. But nothing is *quite* that simple, either, and there are plenty of things it is worth knowing about if you want to get the very best from your track days.

With the above in mind I've tried to explain as much as I can about the track day world, covering everything you need to do to get out on track if you're new to track days, but also including much that should even be of use to seasoned track day goers.

Hopefully, while reading this book, you will learn that there's a degree of balance needed when you're driving a car fast on the race circuit, and the same was true when it came to deciding what to include in *Track Day Manual*. You see, track day drivers are a diverse bunch, and at a typical day you meet all sorts. There are those who like nothing more than tinkering with their cars and then going out to see if they've improved them, while others are only interested in the driving, and others still get as much from the social side of

track days as they do from the action. Obviously then, it would be difficult to please all the people all the time.

For instance, many might question the placing of the preparation and modification stuff *after* the driving technique and the emphasis on the latter. But there are two good reasons for this. Firstly, although *you* might want to modify your car to make it better for the track, it is not essential, and many don't. And secondly, and far more importantly, there is much more to be gained from improving your driving technique than modifying your car to begin with. And that's a fact.

That said, this is not a driving manual, and in the space allowed I could only cover the basics, and in some places this means there may be more 'hows' than 'whys', so apologies for that. I've also tried to avoid the mathematical formulae that litter some race driving manuals – but if you feel the need to delve a little deeper then check out Further Reading in the *Contacts* section (Appendix 2). While we're on the subject of the *Contacts* section, if I have written about companies, products or organisations in the body copy then turn to the back to find details.

So, all you need do now is read *Track Day Manual*, then book your first day and get yourself out on track. See you out there…

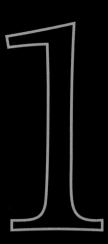

What is a track day?

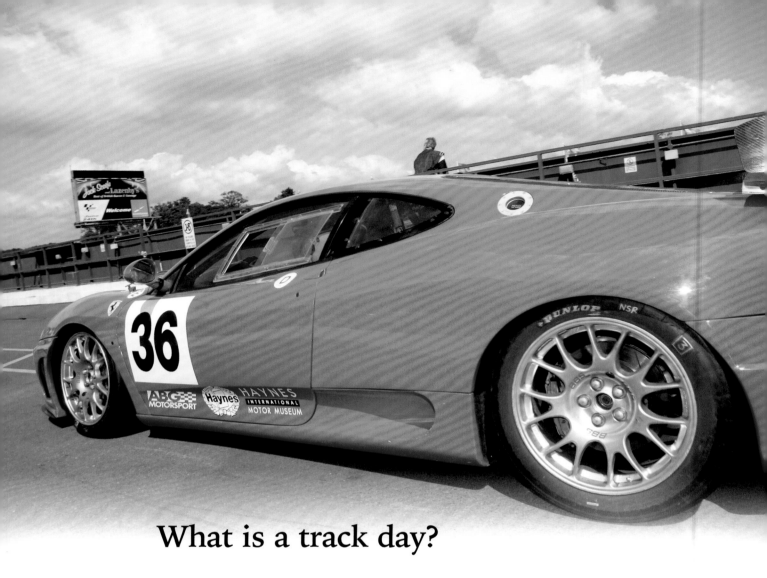

What is a track day?

You don't need a supercar, the skills of a racing driver, or a huge pile of money to go on a track day. But you do have to go along with the right attitude – and to do that you definitely need to know just what a track day is.

I was once told never to start a piece of writing with a dictionary definition. It's good advice, and particularly so, it seems, when it comes to the subject of track days, because the first dictionary definition I found was wrong! Hardly surprising, to be fair, because there's a little bit of confusion when it comes to what a track day actually is and just what sort of people go on them. Some of the blame for this has to be laid firmly at the door of the media, particularly those TV shows which are forever referring to track days as ideal environments for the next new supercar or manic bug-eyed special. The problem with this is that many people then think you can only go on a track day if you have such a car. And that's simply not the case.

The truth of the matter is that there's a track day out there to suit just about everyone and even every car, and while you'll almost certainly get more from the day in a performance car of some sort, there's nothing to stop you taking a regular saloon or estate car out on track.

It's simple, too. In the majority of cases (as I said in my introduction): all you'll need is a current driving licence, a helmet, and a car – and you're away. The only thing you might have to watch out for is an age restriction which, if there is one, will probably require you to be over 18, or on rare occasions over 21. Track days don't need to cost the earth, either, and you'll certainly find one for less than £100 if you shop around, which has to be good value for what is a whole day of real adventure.

But while they may be a fantastic way to get relatively cheap thrills there's one thing you need to know about track days from the very start, and I make no apologies for having this printed in big bold capital letters, as it's very, very important:

A TRACK DAY IS NOT A RACE

Right, we've got that out of the way early, but you can bet it won't be the last time I mention it, since it's only because track days are strictly non-competitive that they're able to take place at all. This is all to do with the personal liability insurance the organisers take out, and it's why they tend to come down hard on anyone caught racing out on track, or indeed timing their laps, but more on that a little later. First, what about where we came in? What would be a correct dictionary definition? Well, here's what Bressy's Concise (so concise this is the only entry) has to say on the matter:

Track day (noun): *a strictly non-competitive day at a race circuit where participants are allowed to drive their cars as quickly as they damn well please.* See also: 'heaven'.

Where it all began

While one-make car clubs have been hiring tracks for many years – and there have also been manufacturer-run days – companies offering track days to individuals in any type of car first began to appear in the 1980s. These were usually clubs and were rather 'high-end' when it came to the sort of cars used. During the 1990s, however, other companies emerged which catered for a broader market – not just restricted to those who were part of a club – and this was the catalyst for an explosion in popularity for the activity.

The growth in the market also resulted from an increasing availability of tracks as their owners came to realise that by hiring out their circuits for track days between testing and racing events they could earn some very useful additional income for their businesses. Consequently, circuits are now far busier than they ever were, and many more people are able to enjoy the challenge of driving

their cars at speed. Indeed, by the end of the '90s track day driving was seen as the fastest growing leisure pursuit in the United Kingdom.

It's not just the UK, though – yet it's fair to say that Britain enjoys one of the most vibrant and diverse track day scenes; arguably because it also has the most traffic-choked and harshly-policed roads. Track days now take place across Europe and the rest of the world, but they're not always called track days ('lapping' is a term used in the US, for instance). It seems that wherever there are cars there are people who want to drive them fast in a hassle-free and safe environment.

Choosing your track day

Google 'track days' and you'll likely find a whole range of sites offering you drives in Ferraris, single-seaters and even Chieftain tanks. These are actually experience days rather than track days (mind you, a Chieftain tank might have been useful at some days I've been on!) and not really within the scope of this book. They can be rigidly controlled, so they lack the freedom of a proper track day – well, would you let *you* loose in a Ferrari? – but if you're given one for a birthday present, then it's a good introduction to the track, and you'll almost certainly get a little bit of coaching out of the day at the very least.

But, assuming you've your own car and you want to do more than just say, 'I've driven a Ferrari don't y'know', then you need to delve a little deeper. A good place to start would be the lists at the back of this book – which are certainly

LEFT You don't need a Ferrari to go on a track day – but if you do happen to have a Ferrari you're never going to get the most out of it unless you take it on a track.

BELOW Empty stands: but the circuit owners can still make money from allowing track days during the week.

ABOVE It's worth asking how many cars are allowed out on track at any one time. Motor Sport Vision at Brands Hatch have the balance between track time and the number of cars allowed out just about right.

the closest source to you at this moment – might include www.pistonheads.com, which has an excellent track day forum, while some magazines will publish a list of upcoming days. You might also try just ringing up your local circuit to see who is running a day there soon (see Appendix 1).

Also, if you're based in the UK, check out the website of the Association of Track Day Organisers, which has a full list of all affiliated companies. Choosing an ATDO-affiliated operator makes a great deal of sense, and not just for your first day, because to be accepted into the ATDO an operator has to conform to a certain code of practice, and if it doesn't comply with this code

it will get chucked out, so there's a good quality-check in place there.

That said, not all of the good track day operators necessarily belong to the ATDO. For instance, the in-house circuit companies will often not belong to it simply because they don't need to. Usually it's the circuits that insist a company belongs to the ATDO before allowing them to run a day at their track, and that's one of the main reasons an operator will join. Obviously, then, if a circuit is running its own days on its own tracks, then it needn't worry about this.

So, although membership of the ATDO is a good starting point if you're looking for an operator, non-membership doesn't necessarily reflect on the ability of the company to run a good day. With that in mind, sometimes it might be worth having a list of questions drawn up before you actually decide to go on a day, to give you an idea of just how well run it will be.

The things to ask about include how many cars will be permitted to attend and how many of these will be allowed out on track at one time, as this will impact on the amount of track time you get. The limits on the number of cars allowed out will vary from track to track, often based on those set by the motorsport governing body, but sometimes they're much stricter because of noise issues.

Clearly, then, if there are 100 cars signed up for the day, but only 15 cars are allowed on track at any one time, then you're not going to get a great deal of action. Incidentally, at this stage you

RIGHT Try to go along to a track day to get an idea of what it's all about before you book your own. They can be fun to watch and there will be plenty of drivers in the paddock who will happily give you advice.

should also check to see if it's an open pit lane or a session day (see below) as this is also likely to have an impact on how much time you get out on the track.

Other things worth checking up on include whether there's tuition available – if you're new to track days you really should get some – and how much it costs. Also, ask whether the instructors are ARDS (Association of Racing Driver Schools)-approved. If they are, you can be sure of their credentials.

If you want a souvenir of the day you could ask whether there will be a photographer present. There usually is, and sometimes they will even have pictures from the very same day on display in the pits, thanks to the wonders of digital photography.

Of course, while asking these questions is all well and good, and certainly advisable, for most the decision will be made by the weight of their wallet. Still, it is worth being a bit careful about just choosing the cheapest day you can find, for as with many other things in life, cheapest does not always mean best. That said, there are some very good budget track day operators around.

If you can, it might actually be worth going along to a track day to see what's happening before you sign up yourself. They're usually free to attend – though check first with the organisers as some airfield days on military bases are tightly restricted – and they can often be quite entertaining as well. By being there, not only will you be able to judge just how well the day

is organised but you'll also be able to chat with people out on the day for tips on car preparation and the like, and for their opinion of the operator and the track day.

Incidentally, if you're looking to get a good deal on a track day, then it's worth thinking about clubbing together with a group of mates. If you approach an operator with the promise of five or six cars attending, there's a very good chance you'll be able to negotiate a discount. Another thing that's worth bearing in mind is that track days can be quite a bit cheaper in the winter, so shop around and you might be surprised at what bargains you can find.

ABOVE On an airfield day, on the other hand, you might find more down to earth machinery. Here a Mondeo follows an Astra.

LEFT Because of financial realities, track day drivers will find their own level when it comes to choosing days, and you'll often find that the expensive machinery tends to go out on the more expensive days. Here a Porsche GT3 heads a brace of Radicals.

ABOVE Donington Park. In 1993 Ayrton Senna made F1 history here, but these guys are driving it in their road cars at a track day. It's a bit like having a kick-about with your mates at Wembley Stadium, when you think about it.

Choosing a venue

It's a fact of track day life that most people want to drive around somewhere like Silverstone or Spa rather than a smaller circuit no one's heard of. There's something about following in the wheel tracks of the greats that has enormous appeal. What sounds more impressive in the bar, after all: 'I've been lapping Spa,' or 'I've been on a track day at Chipping Teacup aerodrome,'? No contest in the kudos stakes there. But the problem with the bigger venues is that track days held at them are usually more expensive, and not simply because of supply and demand economics, but rather because they will often cost much more for the operator to hire when compared to a lesser-known venue. The top tracks do have many major advantages, though, such as much better facilities like cafés or restaurants and the availability of fuel, and safety is usually pretty good, with lots of run-off areas and nice soft gravel traps to stop you if you forget your Peugeot 205 hasn't quite got the downforce or braking ability of a McLaren MP4-23.

Yet when a track day could easily cost as much as £300, and even more at one of the more glamorous venues in the summer, it's easy to see why many will stay away. But, in a way, this is how the track day industry works. There are certain levels of price and everyone tends to find their own level because of the realities of their economic situation.

It doesn't really matter though, there are plenty of other venues out there and when it comes to fun there is no doubt that you can have just as much at a small venue – maybe more at some of them – as you can at a Formula 1-standard circuit.

One piece of advice I would give when it comes to choosing venues is to visit as many tracks as possible over the years you're taking part in track days. This will enrich your experience immeasurably, and also helps to make you a far better driver.

RIGHT While Cadwell Park might not be the most famous circuit, it's certainly one of the most thrilling.

Open pit lane or sessions?

Track days tend to be run in one of two formats – *open pit lane or session*. Basically, these refer to the time when you're allowed on to the track. An open pit lane means that you can go out on to the track at any time, provided the circuit is open for lapping. That said, it will all depend on how many cars are allowed on a circuit at one time, so you might be queuing in the pit lane before they let you on – though this is seldom for very long on a well-run day.

When a day is run in sessions, on the other hand, you'll take to the track at set times for a set period – usually 20 minutes or half an hour – throughout the day. Groups tend to be made up either of those in a similar experience and skill bracket – Novice, Intermediate and Expert is a common way of doing it – or, more rarely, by the performance capabilities of the cars. Sometimes the operator will decide which group you go into; sometimes it's up to you to honestly assess your own ability.

Drivers will usually be given wristbands showing which group they're in, but with many operators it's possible to change group during the day. The need to change can occur when newcomers realise there's a lot more to this circuit driving than first met the eye and that they're not Expert after all. Then again, someone who's signed up for the Novice group might easily find that they're being held up by other drivers in the group, so a switch to Intermediate might make sense.

Both formats have their fans and their detractors, but either way is good and it's certainly worth trying each format before deciding which suits you the better. Open pit lanes allow more freedom and, possibly, more track time in a day, but there's also the likelihood of more of a speed and skill differential between the cars out on the track. As for session days, well in theory there should be less of a problem passing other cars, but you do need to be a lot more organised to make sure you're ready in the pit lane when the session begins.

ABOVE Cars are split into various groups on session days.

LEFT On session days you need to be a bit organised to make sure you're in the line to go out when it comes to your turn on track.

Passengers

There's one thing about track days which you can never get in racing, and that's the chance of sharing the experience, of having someone sitting alongside you in the passenger seat. Please note, though, that's only in the front seat – back seat passengers are almost always not allowed.

Passengers can sometimes go along for free – although they will need to sign an indemnity and be issued with a wristband – but often it will cost a small amount depending on the operator. They will usually also have to be over 16, while under 18s may need parental permission. Passengers will also have to wear a crash hat, which can usually be hired from the operator for a small cost if they don't have one of their own.

One of the downsides is that, people being people, there's often a temptation to try to impress – or even terrify – a passenger with your steering wheel wizardry. Resist it, and remember that by taking someone else out in a car on a race track you're also taking on a very large responsibility.

It's rare, but sometimes track days will not allow passengers, so if you want someone to ride shotgun and you haven't been with a particular operator before, then check first.

The issues

There's no getting away from it, telling someone you like to drive fast is not the way to win friends and influence people these days. To be fair, that's understandable. After all, when people are being told to cycle to work to help save the environment it's hard to feel like a saint when you're tearing

RIGHT Most track days will allow you to take passengers out – but they need to sign an indemnity and wear a crash helmet ... 'Hold on tight!'

around a race track all day. That said, motorsport is beginning to take on board the green message, and in the years to come some of the changes we are beginning to see in racing, such as bio-ethanol fuel, will begin to filter down to the track day scene. Who knows, in a few years time we may all be lapping in hybrids…

Yet it's fair to say that in the grand scheme of things the carbon footprint of track days is minimal, and perhaps it should be offset against the fact that they can provide some very real benefits to the community at large. You see, track days give the owners of powerful cars the opportunity to drive them as they were meant to be driven, thereby – hopefully – removing the temptation for them to do so on public roads. They also offer a safe environment in which people can learn to handle powerful, and not so powerful, cars – which must surely help to cut down accidents on the streets? Indeed, there's a lot to be said for the argument that young drivers should attend track days and take the tuition on offer quite soon after they pass their test.

Noise nuisance

While you might not have much sympathy with someone who moves into a house next to a famous race circuit that's been there for 60 years and then complains about the noise – it happens more than you would believe – you would have to agree that in the current anti-car climate they are going to be listened to. And this is why noise is such an issue on track days.

You'll get *noisy* days, usually at the bigger international venues, but the vast majority of days will be strictly policed, usually with noise tests in the paddock as well as extra tests when the cars are out on track. It's up to you to make sure your car is within the noise limit or, quite simply, you won't be allowed to run. Most regular road cars (even high-performance models) will pass the test easily, but you may have a problem if you've a real screamer or you've fitted an aftermarket exhaust system that ups the decibels substantially.

The lower limit will probably be around the 98dB mark, which is actually easily attainable for most cars, while some tracks go as high as 105dB. Checking before the event can be a problem, but a garage that does government roadworthiness tests – such as the MoT in the UK – may have the necessary equipment. Also, many track day organisers, and even many of the circuits, will happily test your car for free if you turn up on a track day or test day you're not taking part in and ask them to do it for you.

Another noise issue that is just starting to raise

its ugly head is tyre squeal, which really is a pain. So, if you're asked to calm down your driving at certain venues, it may well be down to the noise your tyres are making rather than any problem with your car control..

Timing

However much you tell people they mustn't time their laps at track days, there still seem to be those who just won't listen. Even worse, there are even some who will go so far as posting their times on internet forums. This might seem harmless enough, but if you're claiming on your insurance from that very same day and someone at the insurance company sees the forum, then you can forget it. The insurance for the whole day will be null and void – and that includes the public liability insurance the operator takes out.

It's a serious issue then, and one that's harder and harder to police. After all, some cars have timing facilities built into them these days. The important thing here is attitude. A track day is not a race, it's not a time trial and it's not a test session; it's a bit of non-competitive fun. So remember that, and please leave the stopwatch, and your competitive streak, at home.

ABOVE Track days offer driving enthusiasts a safe environment to push their cars to the limit – *yee ha!*

BELOW A Honda Civic Type R undergoing a noise test at Brands Hatch. Noise is a massive issue in the track day industry and if there are restrictions on your day and your car is too loud then quite simply you won't be allowed to lap.

The track day world

The track day world

There's more than one way to get track thrills in your road car, and while many will choose to sharpen their skills within the wide open expanses of an airfield, others like nothing more than getting it sideways on a drift day. Meanwhile, for the truly adventurous, there's always the mighty Nürburgring.

Airfield days

If you're new to track days then you might want to consider taking part in an airfield day before you take to the race circuits. There are a number of reasons for this, but the main one is the space available. There tends to be very little to hit at an airfield, so there's plenty of room to make mistakes and learn from them.

That said, airfield days are not just for novices, and there's a growing band of seasoned track day goers who enjoy nothing more than being let loose in the wide open expanses of an airfield without the worry of coming into contact with a barrier, tyre wall, or solid pieces of scenery. Indeed, it's not unknown for professional racing drivers to spend a day at an airfield in order to hone their skills – including a few F1 drivers who are regularly coached at Bruntingthorpe in a standard road car, believe it or not.

But just because there's a lot of room, it doesn't mean you can go totally bonkers on an airfield day, as most are well run and stick to the same rules as other track days. Because, even though there's plenty of space, there are still other cars to hit and there's always the risk of rolling a car if you get it very wrong.

Still, airfield days tend to be a little more relaxed than circuit days – in my experience at least – with usually open pit lanes and, because there's plenty of room out on track, less time queuing. There are also fewer stoppages for things like pulling cars out of gravel traps. Even so, you can bet the day will be stopped every now and then for the bollards to be replaced after someone accidentally rearranges the shape of the track by scattering them all over the airfield.

Bollards, or cones, usually a smaller version of the type you'll see marking out contraflows on the motorway (or on top of a statue of a long-dead dignitary if you live in a university town) are pretty much a necessary evil when it comes to airfield days. These will mark the course, but they're easily moved if a car brushes past them, so you need to be ready for this. Personally, I believe the odd rogue bollard can actually liven up a session; it's something to react to, adapt to, and that can only sharpen your driving skills. However, be careful about hitting them, especially the bigger ones, since if you catch them at the wrong angle they can dent your car.

The great thing about marking out the course with bollards is the flexibility it gives the operator to design fun layouts – although the lack of gradient means you're never going to get the challenge or the thrill of a full-on road course. Some of the operators will tend to change the layout on every visit, while others will keep the same layout, giving it more of a race track feel. Some will even add overtaking lanes for faster cars if they've a particularly wide runway to play with, with a long line of bollards to mark them out (this works very well). Sometimes an operator will run two track layouts on the same day, one a slow and tight circuit, the other a faster flowing affair – the possibilities are endless.

Funnily enough, when you think about it, you're not really straying too far from the spirit of many circuits by taking to the airfields, particularly in the UK. After the Second World War, with Brooklands and Donington (the pre-eminent pre-war tracks) out of action, racing had to start from scratch, and the wide, sealed surfaces of bomber and fighter runways left over from the war were ideal. So, many of the top circuits in the UK were airfield venues to begin with. Indeed, at the start of the Formula 1 World Championship, the likes of Fangio and Ascari were wrestling their big front-engined Ferraris and Alfas around a Silverstone airfield that was marked out with steel barrels. And if it's good enough for them …

There are some drawbacks to airfield days, though. For example, noise is often more of an

issue than you would expect from a place that's been used to the rumbling of heavy bombers and the low passes of jet fighters. But many of the airfields used are not now active air bases – which also means you need not worry about being out-braked by an F16 into the hairpin! In fact, in many cases these airfields are now army bases with family accommodation close by. So noise is actually something the operators have to be very strict on, and usually the limit will be around 98dB.

Other drawbacks include the bollards, as mentioned above, and some airfields – though by no means all – can be a bit bumpy, as the surface may be old and worn out, while one airfield in particular seems to be very abrasive when it comes to tyre wear. Naturally, the facilities are not usually up to the standard you would find at a permanent venue either – there's usually just a burger van and temporary toilets – and you'll almost certainly have to leave the venue to find fuel.

Worst of all, though, especially in winter, it can be very, very cold on a windswept airfield, and very wet when it rains. So make sure you take plenty of waterproof clothing, and bear in mind that any gear and tools you take along with you might need to be covered when you're out on track, so a tarpaulin of some sort is useful if it's

ABOVE The circuit will usually be marked out with cones at airfields.

FAR LEFT There's usually plenty of space at an airfield venue.

ABOVE Facilities tend not to be quite as good as they are at permanent race circuits – but the bacon butties at Elvington are top notch!

just your car and you.

On a more positive note, airfield days tend to be less expensive, and if you're insuring your car for the day (see Chapter 10) then the premiums can be cheaper too. So, while Hullavington, Elvington or Colerne might not sound quite as glamorous as Silverstone and Spa, they could be the ideal places to get your first taste of track day driving.

Novice days

Another good starting point would be to sign up for a novice day which – as the name suggests – are track days for novices only. A few operators run these, including Motor Sport Vision at its well-known venues, and they make a great deal of

sense, particularly as some can be put off track days at the very first visit because of the speed others are flashing past them. With the novice days everyone is in the same boat, and there are also plenty of instructors on hand to help out.

Track evenings

Another variation on the theme is track evenings, which take place at a few of the main venues during the summer months. Naturally, there isn't as much track time available as there is on a full day, but they're the perfect solution for busy people who can't easily get time off during the week – and sadly the vast majority of track days take place during the week.

While we're on the subject of when track days take place, it's worth mentioning the track day season, although this has tended to lose its relevance over the past few years as track day operators take advantage of milder winters to put on days the whole year round. Indeed, you'll have little difficulty finding a track day to attend at any time of the year – in the UK at least. That said, the majority will still take place during the recognised racing season, from March until November.

Restricted days

For some, an introduction to track days will come through a love of a certain make of car rather than any real desire to drive fast. Club days on tracks were among the very first track days, and many

RIGHT With no pit garages at airfields you'll be out in the elements. On this particular day the guy who owned this gorgeous supercharged MX-5 was probably glad he'd fitted the hardtop.

still prefer to get out on track with like-minded people in similar cars. For that reason these days can have a good social side to them and they're also often tied in to certain internet forums.

One advantage of these days is that the disparity in speed between the cars is not so great, but this is not always the case – after all, a Porsche day could easily encompass anything from a 924 to a GT2, for instance.

In the United States the track day scene seems to be mostly about days restricted to one marque or another, but that is beginning to change.

Broadening the scope a little, there are also days that cater for just Japanese cars, or just German cars, while there are also now plans for days just for older classic cars.

But even days restricted to one marque have been known to allow the odd interloper in, so if you're keen on doing a day at a certain circuit at a certain time and that day is supposed to be restricted to a certain type of car, it might still be worth getting in touch. They might have one last place they want to fill, and they will usually let just about anything run if it helps pay the bills and it passes the noise test.

Track days abroad

There's no better adventure than taking in a track day abroad, and while it might be difficult dealing with an operator in another country – but less and less of a problem thanks to the internet – you'll be

surprised how many domestic track day outfits will actually organise trips to far-flung venues. Some of these can take place at the major tracks like Spa and the Nürburgring, which can be a bit on the expensive side, but there are also the smaller circuits which can make for good budget choices. Folembray and Croix-en-Tenois, both close enough to Calais in France, are quite popular for this reason.

The obvious problem with foreign trips is getting home if it all goes wrong, and because of this many will trailer their cars when venturing overseas. This does take a little of the adventure out of it, and an ingenious way around it – assuming you're going in a group – is for one

ABOVE Track evenings are popular during the summer months. (*Bresmedia*)

BELOW A gaggle of Porsches at Oulton Park – you'll find that some days are restricted to just one make, or even type, of car.

RIGHT *Le pits*:
Folembray in France,
there's nothing to
match combining travel
with a track day.

BELOW RPM's
Motorsport Tours
were great fun – let's
hope someone else
organises something
similar soon.
(*RPM Promotions*)

member of the group to bring a trailer so that any broken down or shunted car can be rescued.

The major breakdown recovery companies will have policies for those venturing abroad, but it's unlikely they will pick up a crashed car, and some will not cover trailers, so check the small print before signing up. There are plenty of stories of drivers having to arrange recovery home after an off on some far-flung corner of a foreign track. It can be very expensive, and it only becomes a funny story after a few years and a few beers.

Often, for some strange reason, track days at the bigger overseas venues tend to take place just before or just after international motorsport events, so finding hotel space can be a problem as demand is high. So try to book early, and check that they have secure and ample parking before you make your reservation, especially if you're taking a trailer.

Obviously, the cost of a foray abroad will be a lot more than that for a day in your own country – because of the extra fuel, hotels, ferry crossings if needed and so on – but the experience of mixing travel with fast lapping is not one to be missed.

Rally days

Other track day formats to look out for include the 'sprint' one-car-at-a-time sort of track day – uncommon but worth a go – and the 'tour', of the sort run by PR company RPM in the early noughties. RPM's Motorsport Tours were a bit like rally track days, with loose – but not rough – stages, controlled-pace road sections and fast lapping at a selection of tracks, all in one action-packed day. They've not been run for a while, though, which is a shame because they were great fun.

Oh, and if anyone else does decide to organise one, then sign me up!

Hill climb schools

If it's a good old-fashioned blatt up a country lane you're after – without the risk of a tractor coming the other way, random cowpats to skid on, and ramblers to upset – then a day at a hill climb school might be for you. There are schools at the Gurston Down, Harewood, Shelsley Walsh and Prescott hill climb venues in the UK that are open to customers with road-going cars, and they offer a great chance to really push your car in a challenging environment, with the added bonus of some valuable driver coaching thrown in.

Hill climb tracks tend to be pretty short, and a run will usually be over in a matter of 40 seconds or so – depending on car and venue – but there's nothing like it for distilling the high-speed driving experience. Also, the close proximity of the banks, trees and barriers tends to focus the mind a bit, as well as adding to the thrill.

For that reason, before you take to the hills, it's worth making sure you know a bit about your car, how it handles, what upsets it, etc. So, maybe think about doing a regular track day or an airfield day beforehand. That said, there's no pressure to go any faster than the speed you feel comfortable with at a hill climb school.

Drift days

If it's total sideways action you're after, then you might want to try a drift day. Drifting, as a sport, involves cars holding crazy angles around small circuits and then being judged on style; it's been described as a sort of motorised figure skating.

There are now a number of schools that will teach you the rudiments of the art, and these – as well as some regular track day operators – have now begun to organise drift days. At a drift day you basically take it in turns to take to a course and drive your car as sideways as possible. You really do need to use a rear-wheel-drive car for this, though – a limited slip differential is useful too – and there's a fair chance you'll get through quite a few tyres. If it ain't smoking, it ain't drifting, seems to be the attitude.

Techniques to get the car sideways include the obvious one of yanking up the handbrake, and also kicking out the clutch to send a shock down the driveline that'll cause the rear end to kick out. Once it's hanging loose you nail the throttle in order to keep as extreme an angle as possible, and then balance it there with subtle modulations of the gas pedal. It's not easy, though, particularly when you need to perform a transition – changing the direction of a slide from one side to the other – but it's really very, very good fun.

If you don't fancy abusing your car and tyres,

a visit to a drift school, where they usually supply both cars and rubber, will give you a taste for it, while also helping you to hone your car control skills for more conventional track days.

Run What Ya Brung

If you simply want to see how fast your car is, then this might be for you. It's pretty much drag racing's version of track days. The one major difference – apart from the baffling lack of corners – is the fact you are timed.

ABOVE Hill climb schools give you the opportunity to try something a bit different in your road car – the car pictured here at Gurston Down is actually a regular hill climber on a test day. (*Simon McBeath/ maximage.co.uk*)

LEFT If you just love to get it as sideways as possible then a drift day, or a visit to a drift school, could be for you. (*Kenny P*)

Famed UK drag strip Santa Pod, and other venues, hold quite a few RWYB events throughout the year, and these are open to all-comers in all cars – they even had an ice cream van run up the strip once! All you need is a driving licence, and a helmet if your car is likely to complete the quarter-mile run in under 12 seconds – which is really going some, actually. But if you're not sure, there's no harm in taking a helmet just in case.

RWYB is pretty cheap, with a small payment – about a third of the cost of a cheap regular track day – buying you as many runs up the strip as you want. That said, how many runs you actually complete depends very much on how busy the day is, and there can be long queues in the summer. Yet, funnily enough, this is part of the fun, as there tends to be a real car meet flavour to these events.

In the winter it's less busy, but there's the increased threat of bad weather, and rain definitely

stops play with RWYB. This is because the strip, at the Pod at least, is a little bit different from your regular length of race track. It's made up of a layer of asphalt, then a layer of rubber, which is then smeared with a coating of a special 'glue' to make for an extremely high-grip surface. So grippy is the glue surface, in fact, that one track worker told me he once pulled the soles off a perfectly good pair of trainers by walking across it. In the rain though, the glue is washed away and grip is pretty much non-existent.

You'll set off in pairs at RWYB events, which injects a bit of competition into the day, each reacting to the traffic light signals of the 'Christmas tree', an initially daunting cluster of illuminations that is actually quite easy to understand.

At the end of your run you collect your timing slip from race control, and this shows your reaction time to the lights and the time it took you to reach various parts of the strip, including the all-important quarter-mile time.

The technique for RWYB is pretty simple, and for quick times often brutal, with talk of holding it at 4,000 revs – or whatever suits your car, which is very much a matter of suck it and see – and side-stepping the clutch. Four-wheel-drive cars, on the other hand, are often pre-loaded on the handbrake – taken to the point of clutch bite then held there on the brake – and then released. As you can imagine, this is not an activity that extends the life of your clutch. Once off the line it's all about quick and sure gearshifts and… well, that's about it really.

RWYB is not for everyone, and if you're really into your driving you'll be gagging for a corner by day's end, but it's definitely a fine way to find out just how quick your car is.

The Nürburgring

Go to just about any track day in Europe and you're guaranteed to see at least one car with a funny little squiggly black loop on the boot. That squiggle is actually a track map of the Nürburgring Nordschleife – but it's also a badge of honour.

The stickers are available off the internet these days, but that's cheating and a guaranteed way to invite hoots of derision from your fellow track drivers. Because, to earn the right to wear the sticker on the boot of your car, you really should do at least one lap of the track, and this means taking a trip to the Eifel region of Germany – but before we get into that, here's a little bit about the place itself.

The Nürburgring Nordschleife is quite simply one of the greatest circuits ever built, a 14-mile ribbon of unadulterated adventure, featuring everything from jumps – if your car's quick enough – to concrete speed bowls and plenty of gradient to make almost every one of its 176 corners (or 73, depending on your definition of a corner!) a real challenge. Up until 1976 it was actually the venue for the German Grand Prix, but even then it was a circuit from a different and more heroic age, and by the early '80s a newer bent-paper-clip sort of track was constructed at its southern end – the

new Nürburgring which is still used for F1 and other international events.

Meanwhile, the old – or 'real' or 'proper' – 'Ring still winds its way through the forests, these days used for madcap German endurance races, car manufacturer testing – they love to say their products are tested on the 'Ring and even set up fake 'spy shots' to prove it – and the odd track day. These track days, which have the full use of the mighty circuit, do tend to be expensive, but they're definitely worth splashing out on.

There's a much cheaper way to get on to the 'Ring in your own car, though – simply turn up for a public day and pay your money. Public days at the Nürburgring are hugely popular, especially with the Brits, who sometimes even seem to outnumber the locals, and they're a great way to get your track thrills. That said, they can be nerve-racking occasions, and with plenty of motorbikes buzzing around the place and locals who know the track like the back of their hand, the disparity in speeds can be huge. In fact, the first time I went round there some years ago was in a little hire car – something the hire companies are on the lookout for these days, apparently – and the only thing we passed was a coach full of tourists!

Clearly then, a public day at the 'Ring is very

ABOVE A Mitsubishi Evo rides the concrete banking of the Karussel at the Nürburgring. (*Dave Woodall*) different from your average well-policed track day, so you need to have your wits about you. Also, you need to be aware that if things go wrong you could be in for a big bill. For a start, there's a fair chance your insurers won't pay out – it may be classed as a public road but most insurance companies don't seem to see it that way – while you'll not only face the bill for fixing your car but also any damage you've done to the Armco, the cost of any money the track has lost because it was closed to recover you, the cost of recovery, the cost of garaging your car while you figure out how you're going to get it home, and so on.

And things do go wrong at the 'Ring, especially when it's wet. I once did three laps on one very soggy day, and on each lap there was a car in the barriers. But, in spite of everything, it's still worth the risk – for there's no greater challenge in the track driving world. Just remember that the 'Ring takes time to learn, and there are many who have driven it hundreds of times and still wouldn't honestly profess to *know* it. Some even say that the most dangerous time for a 'Ring driver is when they *think* they know it, for it's then that they find out that they don't, and that can be costly.

Many reckon the best way to learn the Nürburgring is to break it down into sections, maybe concentrate on a sequence of corners then give the car – and yourself – a break for the next bit. Some even advise that you should treat it as a fast A-road to begin with, and that's a very sensible approach, too. There are ways you can get a bit of a head start on learning the 'Ring, though, the best-known of which are the driving

games and the in-car camera DVDs and laps posted on various internet sites. But be careful with all these. However realistic they seem, they will not get the gradient right, video just doesn't seem to be able to do that, and that's a large part of the majesty of this wonderful circuit.

Public days at the Nürburgring are not actually track days – although, as mentioned, some track day operators do run regular days there – but are rather the times when the circuit is open to pretty much anything on wheels. Now these open days are actually not as regular as they once were, and often the only times that the public can get on the 'Ring are for a few hours during the evening, such is the demand for its use by motor manufacturers. So don't just turn up and hope to be allowed in; check the opening times on the internet first – full

contact details in Appendix 2 – before you set off to drive to Germany.

To get out on track is simplicity itself: you just buy a ticket from a machine or the little ticket office and slot it into the machine at the entrance to the track, the barrier rises and away you go – just like a multi-storey car park. If you're going to do a lot of lapping then it works out cheaper to buy a multi-lap ticket, otherwise it can prove expensive (it's about £13 a lap at the time of writing).

One last thing. The Nürburgring is not to be taken lightly – it can bite. But because of this it remains one of the few great driving challenges. Indeed, it would be a terrible shame if any of the all-too-regular rumours actually turned out to be true and it was closed to the public. Get there, before it's too late.

BELOW As you can see, in some places the barriers can be very close at the 'Ring. (*Dave Woodall*)

Track day cars

Track day cars

You can take just about any car on a track day, but some will naturally be better than others, and while many will track day their daily rides, others will want to use cars that will only get driven on the circuit – which all goes towards making for a truly diverse track day scene.

Chances are you've already got the car you want to use on track. It's your daily driver, perhaps, the car you use for work and shopping. But maybe it's got a little extra something, maybe it's from the sportier end of the product range. If so, that's a good starting point, as the vast majority of cars you'll see out on a track day are a little warmer than your standard fare.

But that's not to say you need to rule out the idea of going on a track day if you've something a little bit on the ordinary side. If you think about it, these days a regular hatch or saloon will often out-perform a sports car from the '70s anyway, and you'll still see plenty of enthusiasts out on track days in the older cars. So, pretty much anything goes. Indeed, you'll be surprised at what can turn up at track days. I even saw a London taxi out at Donington once – honest!

The point is, as long as it's safe and sound you should be OK. Some track days will ask that the car has a current roadworthiness certificate – the MoT in the UK – but that's rare and usually it will be up to you to ensure your car is safe. Indeed, there's rarely any sort of scrutineering at track days, so the onus is entirely on you to make sure everything is working as it should and that the car is not losing oil, water or wheels.

The car should really be your own vehicle, too, or at least one you've permission to take on track. You probably won't be too surprised to read that people have been known to take hire cars out on track days. But be warned, the hire companies are getting wise to this – especially when it comes to German companies and the Nürburgring – so bear that in mind, and remember that should you crash it, you'll not be insured. Not sure what happens then, but it's not going to be pretty and it will almost certainly involve a solicitor or two.

But if you're stuck for a track ride, the good news is you don't need to spend a fortune to buy a car that's not just adequate for track use, but will also be damned good fun to boot, as there are plenty of cheap and quick-enough cars on the second-hand market. How cheap? Well, to give you an example of what's possible, some years ago the sadly missed *Cars and Car Conversions* magazine ran a feature that's been much imitated since. With a budget of £500, four of us staffers were to buy a car for track days and then we were to meet up at Brands Hatch to see which one was the best all-round package. The idea was to take them 'as seen', with no modifications, so it was pretty much a thorough safety check then straight to the track.

It was actually quite a high-end track day, so the 'bangers' were out with everything from a pukka Ford GT40, all types of Porsches, a couple of Ferraris, and the usual smattering of Lotus Elises and Caterhams.

Amazingly, our mixed bag of old BMW 325i, Metro GTI, Citroën BX and Astra GTE did not look out of place in such exalted company and we even had a few cars having to move over to let us through. Then Touring Car driver Phil Bennett was on hand to assess the cars, and the picture of him in the Metro in front of a brand new Subaru Impreza WRX – the driver shaking his fist at the ignominy of being overtaken – was something to savour. Of course, much of this was down to the driver, and it's a fact that to begin with there's a lot more speed to find with driving technique than modifications. But still, the main point is clear: you don't have to spend a fortune to pick up a perfectly adequate track car.

The Metro was actually the star of the show, so much so I kept hold of it and used it as my own track day car for the next two years. I always meant to modify it, but never had the time, and so

it never really cost me much more than the original £500, and I did quite a few days in it in the end without any real problems.

I even believe it can actually be more fun driving a cheap and relatively slow car than a mega-bucks supercar, but then that depends on whether you're after smiles or thrills, I guess. Personally, although I'd quite happily make room on the driveway for a Ferrari 430 Scuderia, I'm also very happy with my near-standard (just fast-road brake pads and better discs) Mazda MX-5 for track day fun. It's cheap, near bullet-proof, and guaranteed to put a smile on your face.

Buying for the track

Yet while it is perfectly viable to buy a cheap car for track days, some of the problems that we actually encountered with the cars during that *CCC* feature are worth bearing in mind. For instance, the old Astra was fine on the road but once out on track it did an extremely good impression of a cloud factory and was actually black-flagged before it could get any serious laps under its fan belt. It turned out there was a real problem with oil surge, on a scale that is pretty unusual, and a baffled sump would have to be fitted if it was ever to be taken on track again.

As I said, that was unusual. But still, just how can you be sure a car might not have a problem like this *before* you get it out on track? The best way is to do a bit of research before you choose your track day weapon. First of all, go to a couple of track days – you shouldn't have to pay to spectate – and there will be plenty of people in the

FAR LEFT Standard MX-5 and Ford Mondeo in action at Elvington – there's nothing to say you *have* to modify your track car.

BELOW The author's old Metro GTI (rest in pieces) cost just £500, yet was still great fun out on track. (*RPM Promotions*)

ABOVE If you're thinking of buying a car for the track, then take a trip to a track day first and talk to those who are running the sort of car you're after – most will be happy enough to talk about their cars.

paddock more than happy to talk about their cars. If you have your eye on a particular model then chat to someone who's running one, and if you can't find anyone who is, then ask yourself why. Maybe that car is just more hassle than it's worth?

Similarly, check out what cars do seem popular. There are reasons why you'll invariably find certain cars at just about every track day: some are just great fun, while others are readily modified, and others still give great bang-for-your-buck value. The thing is, if you choose a car that's popular

there will be more people to help you, more advice on the internet forums and more likelihood that the problems you'll encounter will have already been solved by someone else.

There's even an argument for buying a car that's been used on track already, especially if it's been developed and beefed up for the job; but that really depends on what you want. If you're looking for a car to use on the road, in comfort, as well as the track, then just find one the usual way, and check all the things you would normally check when you're buying a used car – service history, evidence of crash damage, the chassis numbers, and all the other must-dos that are beyond the scope of this book but widely available elsewhere.

Don't automatically go for something just because it's cheap, either. Book price is just the start of it. If it's budget thrills you're after, then you need to look at the cost of ownership, always remembering that the car's going to be put through a lot more than the manufacturer had in mind. Tyre and brake wear tend to be the real problems, so perhaps look for a lightweight car if you can, as these are usually less demanding on pads and rubber, and then there's spares to think about. You may well need them, and for some cars they can be expensive and hard to source.

Wear and tear

One thing you should bear in mind is if the car you buy – or indeed the car you already have – is going to double up as road and track car, then it's going to suffer for its art. There's no getting away from it, the track is a far more hostile environment

RIGHT Your car will never be driven as hard on the road as it will be on the track.

than the road. One track day preparation specialist even told me that one track mile is equivalent to ten road miles in terms of the wear and tear on the machinery.

With that in mind, you'll almost certainly be getting through more oil, so you'll need to monitor the level regularly, which includes during the course of a track day. And think about changing the oil far more regularly than normal, too. Try to keep an eye on your track mileage, and base your service and oil changes on that – always keeping in mind the one-to-ten track/road ratio. You'll also need to look out for the increased wear rate of brake pads and tyres in particular. Remember, your car will never be driven as hard on the road as it is on the track; even if you do hold the record for the supermarket run…

If you're not really into doing your own work on the car, then maybe take it to a garage. If you can, try to find one that knows at least what a track day is and what it involves – which might be a bit harder than you would expect if my experience is anything to go by – then try to build up a relationship with them. It's easy enough, mechanics tend to be a bit charmed by something out of the ordinary, by the thought of their work being put to the test at Brands Hatch or even the Nürburgring. When you go back they'll want to know how you got on and, who knows, perhaps they'll suggest their own modifications?

Even better would be to find an outfit with a real track day pedigree to look after your car – a company that's an expert in what works well for your particular make. Be aware, though, that many

of these have a motorsport background and can be expensive. But you get what you pay for – so if you want the best…

We'll go into the oily bits in a little more detail later in the book, but for now just remember your car will get a lot more punishment on track. Always bear that in mind, and then there should be no reason why your road car could not quite easily double up as a track car.

That said, there does seem to be a move towards bespoke track day cars these days. Indeed, on some of the high-end track days the majority of the cars will be just for the track, while even lower down the scale there seems to be more and more stripped out and caged-up hatches, some of which are getting close to race-prep specification, and many of which require trailers to get them to the circuit and back.

Trailer or drive?

So, to tow or not to tow, that is the question. Whether it is nobler to drive your car to the circuit or trailer it depends largely on what you want from a track day. Many actually think that the drive there and back is a big part of the day – this writer among them – and if you're going to trailer a car then you might as well go racing.

On the other hand, some cars – especially ex-race cars and track day specials – are simply impractical on the road. Also, there's no doubt that you can push a little harder if you've a trailer – or even a transporter of some sort – to haul the car home should you wreck it or grenade the engine. So there's the peace of mind argument, too.

BELOW Some will trailer their cars to track days.

ABOVE A covered trailer has its advantages.

You'll even see quite a few who have excellent track day set-ups of mobile home and trailer, which makes an ideal base for a weekend away complete with a fridge full of cold beers and a barbecue. It's near perfect for a trip to the Nürburgring.

But you need not go to quite those lengths, and a simple trailer, and even a tow car of some sort, shouldn't set you back too much. Of course, the smaller and lighter cars will only need small and light trailers, but if you've something heavier you may have to pay out a bit more. At the time

of writing you could easily pick something up second-hand for under a grand (£s), while at the top end of the market good quality covered trailers will be a lot more expensive.

One thing you'll have to be aware of, though, is that trailers will take up space. It's not as though you can simply park them up on the road outside, after all. Also, a trailer – and a tow car come to that – will need to be maintained if it's going to get your car to track days without incident. Judging by paddock natter, some seem to have as many adventures and near misses with their trailers and tow cars on the road as they do with their track cars out on the circuit!

Often trailers will be left out in the open between track days, even over the winter, during which tyres can perish and brakes can seize, so make sure you check your trailer some time before the event, giving yourself time to fix it.

Another thing about trailers is the amount of grit and rubbish that can be thrown up from the rear wheels of the tow car: it can certainly ruin the paintwork on the front of your track car. So maybe you should think about installing a shield of some kind or, better still, investing in a covered trailer – which also gives you somewhere to take cover if it pours down on an airfield day.

When trailering, you must make sure the car is loaded correctly. If you've bought new there should be a manual to show you how; if not, then consult the manufacturer. The rear of your car snaking on a track day can be fun, but a tow car and trailer snaking down a motorway because the latter's unbalanced is just downright scary.

RIGHT The Peugeot 205 may be getting a bit long in the tooth, but its fun lift-off oversteer mannerisms make it a popular choice for the track. (*Bresmedia*)

Which track day car?

There are usually no technical regulations to adhere to on track days, and no classes in which you must run, so it should be no surprise that there can be a bewildering array of cars out on a circuit. It's also part of the appeal, to be honest, for while there is no racing or timing, for reasons we've already gone over and will go over again later – it's that important – it's always good to see how your car stacks up against another.

Yet even though there are no classes, it's still possible to put the cars people tend to use on track days into informal groups, and it's a good way to show you what is available. Now, there's only space to cover some of the favourites, so don't worry if your car's not mentioned, and please don't get upset – there's nothing wrong with standing out from the crowd; and you obviously know something the others don't.

Budget hot hatches

The first of these groups has to be seen as the entry level for track days, though some of these cars are often developed to the stage where they're as quick as race cars around a circuit. If you're on a budget, older hot hatches tend to be cheaper and, when it comes to track day thrills, they're often the better bet, too. They're almost always lighter – which means they can be kinder on the brakes – and they often tend to have much sharper handling than the cars that came along to replace them. Track day hatch favourites include older VW Golfs and Peugeot 205s, both of which offer plenty of tuning options and are fun to drive on track. But whichever budget hatch you choose, you need to check it thoroughly – chances are it will have been driven by someone like you!

Budget saloons

Think rear-wheel-drive saloons; think BMW. There are not many car companies with such a track heritage, and the Bavarians have always made cars for driving. So it will come as no surprise that most track days will feature a few of them – from stripped-out M3s to regular saloons – and they usually make for a cracking track drive. Other popular rear-wheel-drive saloons include Ford Sierra Cosworths and older Fords such as Escorts and Capris. The latter two are great fun, if a bit slow these days unless they're well modified.

As for front-wheel-drive saloons, while a long wheelbase front-driver can push on a bit in the turns, and some can be a little underwhelming on track, they can give you a stable base from which to learn your circuit craft, with less pronounced weight transfer along the car.

Budget sports cars

You might think a rear-wheel-drive sports car is beyond your means, but there are excellent bargains to be had, particularly when it comes to older Japanese models such as the earlier MR2s and MX-5s, which tend to be lightweight and reliable, as well as huge fun to drive. Even Porsches can be cheap and while a 944 may look a little like the unloved 924, out on track it can be fantastic with its characteristic superb balance and good brakes.

ABOVE Anything goes – just make sure you take the dog out of the back first!

BELOW Mid-engined, rear-wheel-drive, and fun – original MR2s can still be picked up for a decent price.

ABOVE Mitsubishi Evos
and Subaru Imprezas
are mainstays of the
track day scene.

Four-wheel-drive

Now, we're not talking here about the Chelsea
tractors that everybody loves to hate, but rather
the Japanese high-performance 4x4 saloons that
have been a large part of the track day scene for
the last decade or so. Step forward the Subaru
Impreza and Mitsubishi Evo, each of which has
a healthy following, while there are also plenty of
bolt-on bits to make them go even faster.

Buying new

If you're buying a new car to double as a road car,
then there are plenty to choose from; although at
the time of writing there does seem to be a trend
towards heavier cars, even with the hatches. Also,
it's getting more and more difficult to circumvent
the driver aids and the stability packages on many

new cars – although these are great for the road
they're not always so good on track.

One thing you should definitely do before
committing yourself to buying a new car for the
track is talk to someone who has driven it on a
circuit – not always easy if it's a brand new model
– or at least read a track test in a well-known
magazine (evo, for instance, which has some very
good track drivers on its staff).

Sports favourites

If there isn't a Caterham or a Lotus Elise present,
then you're probably not on a track day! These two
cars – and, in the case of the Elise, its derivatives
– are massively popular, and for good reason. The
Caterham Se7en, even in its lower-powered entry-
level incarnation, will embarrass the most exotic of

RIGHT New hatches,
such as this Honda
Civic Type R, are big
on performance these
days, and many find
their way on to the
track.

LEFT There's not much that can match a Caterham when it comes to on-track thrills.

supercars through the twisty bits on track, and the more powerful versions are simply mental. Such has been the success of the Caterham concept (which was originally the Lotus 7) that other companies have made similar cars, and often these can be in kit form – Caterhams are also available as kits – with Westfield being the best-known of the Cater-likes.

The Elise has similar track capabilities to the Caterham. It really can feel like driving a race car – with its engine right behind you – and is perhaps a slightly more practical option if you're looking for something for both road and track.

Another popular car is Porsche – the circuit-bred 968 Club Sport is a favourite for good reason; and if you've a bit of money there's always the GT3, GT2, or Ferrari 430 Scuderia, or… well, this particular wish list could go on for ever.

Specials

Track day specials are pretty much cars that are only any use on a track. Even so, manufacturers in some cases will claim they're road cars, too, arguing that they're fun cars in much the same way someone might have a motorbike for thrills rather than for getting from A to B, which I guess is a fair point.

Radical, which is also a very successful manufacturer of race cars, is probably the best known of the special makers and has been a pioneer of motorbike-engined cars, although the wacky 'naked' Aerial Atom now has a very high profile, too. And there are plenty of other, usually small-scale, manufacturers willing to take your money in return for track day thrills, often via mad power-to-weight figures which are sometimes

BELOW Radical's SR3 may look like it belongs on the grid at Le Mans yet – believe it or not – it can actually be used on the road.

thanks to lightweight bike engines. Yet, ironically, one new special maker on the scene is actually a bike-maker using a *car* engine in its amazing carbon-tubbed creation. That's KTM, whose X-Bow looks set to take the track day world by storm at the time of writing.

Racing cars

Seems obvious, doesn't it? If you want a car for the track, then buy a car that's built for the track. It can actually make good financial sense, too, as quite often race championships will die off for various reasons, or perhaps move on to a newer model, leaving cars with nowhere left to race. Because of this you can easily source well-prepared cars for relatively little, usually in the back of the motorsport magazines or, increasingly, on websites such as the hugely popular www.racecarsdirect.com.

There are a few things to look out for, though. For a start the car is unlikely to be road legal – though it might be and rally cars, at least, should be – or easily converted back to road-going spec. So, chances are you'll need to tow it to track days. Also, make sure you get something with a durable engine. Not all race motors are made to last, and the last thing you need if you're after cost-effective action is to be rebuilding your engine after every other track day. Remember, you'll be doing far more laps at a track day than you would at most race days, where 30-minute qualifying sessions and 10-lap races are the norm.

There's one other major consideration before you opt to go down the race car route, and that is that just a few of the operators are unwilling to let racing cars out on track. This is getting rare, though, and indeed, some track days in the UK are beginning to look like test days, which is the cause of some controversy.

Also, if you're flicking through the small ads and you get tempted by a single-seater racing car as your track day weapon of choice then you might want to think again. I say that because most circuits simply don't allow them on track days, mainly because of problems of speed differences and visibility in other track users' mirrors (although the same could be said of many specials, so I've always found this a bit bewildering).

One operator I've spoken to tried to run a session on a day just for single-seaters, but it didn't attract enough cars; which is a shame, because there's nothing quite like a single-seater for kicks, and there are plenty of them out there with no place to race. If you've got one, though, then try Donington's Trakzone days, or maybe other companies running days at Donington, as

at the time of writing it's the only major circuit allowing formula cars out on track days. Other than that, you might venture abroad, as it doesn't seem to be an issue in some other countries.

Syndicates

If you really want to track something a bit special, but just can't raise the readies, then there's always the possibility of sharing the cost by sharing the car. This is getting more and more common at track days, whether it's a couple of blokes building up a track car together and taking it in turns to do a session, or something on a more formal basis, with a group of people owning stakes in the track car – in much the same way as boat- and aeroplane-owning syndicates work. The only difference is that all the members use the car on the same day – although it could work on a

taking it in turns basis, I guess – with sometimes as many as four people hopping in and out of the same car.

That's not as unsatisfactory as it might sound, either, for there's usually plenty of time at a track day and you'll often find you won't want to be out for the entire day anyway. Three would probably be the optimum number, though, simply because there are usually three groups at sessioned days. Of course, this would work even better at an open pit lane event.

A possible problem with a syndicate is who keeps the car when it's not on track. Not so much of an issue if it's not road registered, in which case you'll probably be glad if someone else sorts out the storage, but if it's a cool sports job, then whoever keeps it in between events is going to have lots of fun at your expense. That said, there are ways

BELOW Form a syndicate with a group of mates and you could get yourself a share in something a bit special – who knows, maybe even a Porsche GT3?

around this – perhaps you can take it in turns?

If you are sharing a car, you really should do it with someone you know and trust, a friend would be ideal. Just make sure you all agree on what you want, how long you intend to keep it, how many track days you're going to do, who pays for the damage if it's shunted – the driver who crashes or all of you? – and so on.

One other thing to bear in mind if there are two or three of you sharing the car is the increased wear and tear. Usually there's a certain amount of downtime at a track day, for a solo-driven car, but a car that's on track for the whole day is bound to take more punishment. So you'll pay out more on parts, tyres, brake pads and so on. On the other hand, some track days will charge for just the car, with maybe a small amount for additional drivers, so there are savings to be made there, too.

A syndicate is not for everyone, but if you've got a couple of close mates and between you can just scrape together enough for that Lotus Exige you've all fallen in love with, then why not give it some thought? Just make sure everyone understands exactly what the deal is.

Track day hire

While you would be taking a bit of a risk bringing a regular hire car to a track day, it might surprise you to find out that there are a growing number of companies hiring out performance cars specifically for track days. If you know anything about how the motor racing world works, though, then it shouldn't be that surprising, for race hire is the way most young drivers work their way through the formulas – a sponsor (or dad!) paying for the privilege of having their driver at the wheel.

There are now plenty of companies offering track day hire deals, with a wide range of cars available, from little MX-5s to race-spec sports prototypes. The great thing about this is that you don't have to worry about all the wear and tear and the hassle of looking after the car. The downside is that it can be expensive. Not always, though, and you can get hold of something quite special for a reasonable amount of cash, certainly not as much as you might at first expect. You do need to know if there's any insurance, though. And if so, whether you'll be liable for the excess – you probably will – and how much it is. (See Chapter 10 for more on track day insurance). If there isn't any insurance, then you need to know just what happens if you crash it. It might be a 'you bend it, you mend it' arrangement, or you might be asked to lodge a bond – usually payable by credit card – before the day.

Also, some – but I should stress by no means all – track hire companies will not let you out on the track in the car on your own. This does mean you get a day's coaching, but it also takes some of the freedom from the experience – which to my mind is a big part of the appeal of track days. Sometimes some of these hire cars seem to be going *very* slowly, too, which makes me wonder how much the instructor is allowing the driver to actually push.

That's something you need to discuss with the company before signing up, as is the amount of track time you'll get for your money. Bear in mind that quite often the car will be shared among a group of punters, so the deal might not be as good as it first seemed. If you know what you want, though, and you fancy getting behind the wheel of something a bit special on track, then this could be for you.

On the day

On the day

The really great thing about track days is that they're pretty much free of hassle. That said, there are a few things you'll need to do before you're allowed out on track, and some simple rules to remember when you're lapping.

The wonderful thing about driving *to* a track day is being able to enjoy that very rare feeling of not having to make the best of any sudden thinning in the traffic; of not having to steal your thrills when there's an empty stretch of road, or when there's none of those evil yellow boxes to be seen. Today, none of that matters, because soon you'll have a whole race track to play with.

You'll still need to exercise a little patience when you get to the circuit, though, because there are things you have to do before you hit the track, all of them totally necessary. All operators tend to do things a little differently, but a good organiser should be taking you through a process similar to the one outlined here. The details may differ, but the content should be pretty much the same on whatever day you attend.

The first thing to mention is that it's always worth getting to a track day as early as possible, particularly if there are pit garages and they're allocated on a first-come, first-served basis, as it's definitely better to have the comfort of a garage

rather than slumming it in the paddock, especially if it's raining.

It's actually quite interesting how people tend to park up or choose garages at track days, with those in the same types of cars often gravitating together. It's one of the great unsung attractions of track days, actually, as it's always good to compare notes with someone who owns the same car as you do.

Once you've chosen your spot you'll have to sign on. You'll be asked to show your driving licence at this stage – whatever you do, don't forget it – and in the UK perhaps the green counterfoil, too, although this is rarely the case. Passengers will be required to sign on as well,

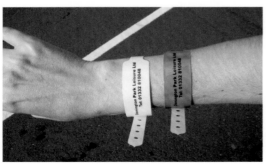

while both drivers and passengers will need to sign an indemnity form which you really should read carefully. Signing this means you understand the risks involved in track driving and that you agree that the operator and the circuit is not liable for any damage you do to yourself or your car.

Once signed on, you're usually given the first of your wristbands, which is pretty much your licence to thrill, and you'll need to show it to the marshal at the end of the pit lane every time you want to take to the track. You might also be given a number to stick on your car – so that marshals can identify the naughty drivers.

Depending on the day, a noise test could be next. This involves bringing your car up to a specified amount of revs, which is usually roughly around three-quarters of maximum revs. While one noise tester will keep an eye on the rev-counter another will hold a decibel meter to each tail pipe – usually about half a metre away from it. Most regular road cars, even high-performance cars, should get through without a problem, even if the level is set at 95dB, which tends to be the lowest you'll find. That said, some bike-engined cars in particular seem to struggle, and those with big bore exhausts that are just there for the noise-effects might have problems, too.

But if you do just scrape through the noise test by a decibel or two, don't think you're in the clear. If the car sounds especially noisy out on track they'll pull you in for another test – that way they can keep an eye on people who modify their cars, with tape and baffling, just to get through, then rip it all off later. It might seem a bit of a pain to you, but remember, all it takes is one noisy car to close down a day and it could even lead to the venue being lost to track days altogether.

ABOVE LEFT Sometimes you can book a pit garage; at other times it's on a first-come, first-served basis, so it pays to get there early.

ABOVE Sometimes you'll be issued with a number.

LEFT Once you've signed on you'll be given the first of your wristbands. You'll usually get the second after the briefing.

ABOVE Unless it's
a 'noisy' day, which
is rare, you'll almost
certainly have to take a
noise test. (*Bresmedia*)

Incidentally, some operators will only monitor
the noise from trackside, so there's a chance
there may not be a formal test, but you'll still be
called in if your car's a screamer.

Once you've passed the noise test you'll
usually get a small sticker to put on your car so
the pit lane marshal will know you're quiet enough.
Then it's time for the drivers' briefing.

Lecture notes

Drivers' briefings at track days are actually quite
interesting from a psychological perspective. Look
around the room and you'll see a few heads go
down – particularly among the newcomers – with
talk of safety and black flags, and what to do in
the event of an accident. What had begun as a
bit of fun suddenly seems deadly serious. You
shouldn't let it put you off, though, because it's
important. The operator is obliged to go through
all these things since there could be massive
insurance implications if it didn't.

The better operators have at least developed
a way of making these briefings a little bit
entertaining without obscuring the underlying
message. So, rather than just saying 'don't do
this, don't do that', a good briefing will include a
few funny anecdotes, too. It's about getting the
balance right, and it's about keeping it as short
as possible.

But just because it's something that has to be
done, it doesn't mean you can ignore it. Some of
the things that you'll be told are vital. For instance,
they'll tell you what side of the track to overtake
on, whether you should indicate for someone to
pass, what the flag signals are, the timing of the
day, and a host of other very important things.
Some of the operators will also run through
particular hazards around the circuit and the state
of the track surface, and they may also take the
opportunity to introduce the instructors.

One last thing on the subject of the briefing –
don't miss it! There may not be another one, and

RIGHT The drivers'
briefing is a vital part of
any track day.

if you haven't got your second armband, which proves you've been to the briefing, you might not be allowed out on track.

On the day preparation

We will discuss what you should be wearing in Chapter 11; so, assuming you're all suited and booted – or at least sleeved and shod – let's have a look at what you should be doing to your car to make sure it's fit for track use on the day.

A great deal of this is actually common sense, but with the same mistakes being made over and over again it makes you realise just how rare a commodity this has become. One of the key things to remember (and the thing that's so frequently forgotten) is that you should always make sure that anything loose is taken out of the car. When a car is on a circuit it will be subjected to forces it will never get near to encountering on the road. Map books will fly off parcel shelves, half-finished packets of mints will become a dangerous distraction in the footwell, and WD40 cans that have been stowed and forgotten underneath the driver's seat will roll into the pedal box and beneath the brake pedal. Sounds unlikely? Well it's happened, and the driver concerned was lucky to get away with it.

According to one track day organiser, there have been even sillier transgressions, including a Porsche 944 with a set of golf clubs, a tennis

ABOVE You need to make sure your mirrors are adjusted correctly.

racket, and a pair of walking boots on the back seat. I kid you not.

Similarly, loose items in the boot can cause real problems, and many a track day driver has picked up a dent – from the inside! – just because a jack has not been secured in place.

You should also check fluid levels – water, oil, brake fluid – all of which we will deal with in more detail in the technical sections later in the book. Many will also go around the car with a torque wrench checking the wheel nuts are properly tightened. This is always a good move, as is removing any wheel trims that are likely to make like a frisbee the first time you take a corner at real speed. It's also absolutely vital that you check your mirrors are adjusted correctly, as these will become some of the most important components on your car during the laps ahead. And make sure the lights and indicators are working as well.

Another thing that you'll need to check, and thereafter keep an eye on throughout the day,

LEFT Checking the tightness of the wheel nuts before you go out on track is always recommended.

BELOW You should always keep an eye on fuel levels – and not all cars have petrol gauges! (*Bresmedia*)

RIGHT Some operators will insist you fit a towing eye, and maybe you'll be asked to tape up the headlights, too.

is the fuel level. This advice might sound a little patronising, but you would be surprised just how many people run out of petrol on a track day. Bear in mind that out on track you'll be constantly in the upper reaches of the rev range, and consequently your fuel usage will be much greater than in normal road driving. If you do run dry it will probably result in the session being stopped while your car is recovered, which means everybody else will lose track time and you won't be the most popular person on the track. Some operators have even fined drivers for running out of juice. It catches out more drivers than you would believe – so keep an eye on that petrol gauge.

If you have a special or a kit car of some sort you might want to be a little bit careful about filling the tank to the brim, though, as some of these cars are notorious for leaking fuel on to the track.

Some track day drivers will take the spare wheels out of the back of the car before going out, too, but this is usually to lighten the car and seems a bit pointless in truth – after all you're not in a race – and although it will undoubtedly shave a few pounds off, it might just take those pounds from somewhere they were needed. Remember, particularly if you've a road-going performance car, manufacturers spend millions to get the balance of their cars just right, so it seems folly to mess around with it too much.

The same goes for tyre pressures. There's probably more talk, and certainly more pseudo-science, concerning tyre pressures at track days than anything else. Check them, of course, but don't stray too far from the manufacturer's recommendations to begin with (see Chapter 12). Also, make sure you always check them with the same gauge, as while you can rarely be 100 per cent accurate with tyre gauges, you can try to be 100 per cent consistent – which is much more important.

Some other things you might be required to do, depending on the operator, is fit a towing eye to the car – on new cars you may have to refer to the handbook to find out where it's kept and how

you screw it in – and maybe tape up your lights, to stop broken glass from littering the circuit in the event of a shunt. But all this should be made clear before you attend.

While we're on the subject of taping things up, you'll find that before going out on track quite a few people will stick tape over their registration plates to obscure the numbers. Some seem to think it's vital, but to be honest I can't see why. It's not as though going on a track day will negate their insurance – because it's unlikely they're insured when they're on track anyway (see Chapter 10). Some say it's because their plates could be cloned, but there seems no more risk of that on track than driving down the street. Of course, if they're in company cars or hire cars, or cars that simply belong to someone else, then that's a different matter entirely!

Incidentally, it's very rare that you will get any form of scrutineering on a track day (although I've heard it can happen in the States) for the

RIGHT It's best to check tyre pressures throughout the day.

LEFT It's always worth taking a good set of tools, brake fluid and oil along with you.

simple reason that if an operator checks out your car to make sure it's safe before a day, and then something terrible should happen, by giving the car the okay the operator has foisted a certain level of culpability on its shoulders. Clearly then, it's up to you to make sure your car's not about to fall to bits, lose a wheel, or leak oil everywhere – both for your safety and the safety of other track users. Note though, that most operators will also reserve the right to kick any obviously dangerous cars off the day.

As to what tools and consumables to take, it depends on how much fiddling you're planning on doing. There's no rule to say you have to take anything at all, but some kit is obviously useful to have to hand. For instance: a good set of spanners, a comprehensive socket set, a torque wrench, a tyre gauge, a compressor or foot pump, tank tape (otherwise know as 'gaffer' or 'crash' tape), tie wraps, jubilee clips, an assortment of nuts and bolts, brake fluid, oil, and water.

A tarpaulin is also useful for covering all this up when you're out on track, especially if there are no pit garages and it's raining.

Easy, tiger!

There's every chance these days that your first taste of the track will be doing some slow familiarisation laps in a long queue of cars – which is sometimes called 'ducks and drakes' – with an instructor leading the crocodile of cars for two or three laps, making sure that everyone knows which way the track goes. One tip here is to try to get to the front of the pit lane immediately after the briefing. The nearer the front you are, the better your view of the instructor's laps, which will help you find the correct line if it's not marked with bollards.

You should also use these familiarisation laps to check where the marshals' posts are, to look out for run-off areas, and to check the state of the track surface. They're reconnaissance laps

BELOW Ducking and driving: two or three 'ducks and drakes' laps will often take place behind a course car before you're let out on your own.

excitement (perhaps understandable when you think they've spent most of their driving life stuck in traffic). The trick is not to let the freedom go to your head, but to build up your speed slowly and (whatever you do) don't feel the need to try to match the speed of the other cars.

This is important. Do not take your ego out for a spin during the early laps of a track day. If a car is faster than you and your car, then just let it go; for right now you need to concentrate on building up the speed slowly. OK, your Porsche might be about to be passed by a Metro, but its driver might also have a great deal of circuit knowledge and oodles of experience, and in this game that can count for a lot.

For most novice track drivers the race circuit is an alien environment – which, incidentally, might account for the looks of some racing drivers – where common reference points such as trees, hedges, side roads and approaching traffic have been banished. Give yourself time to get used to it. And remember, provided you keep an eye on your mirrors and stick to the right on the straights (if the overtaking is on the left on that particular day), then no one is going to mind you trundling along at whatever speed you choose.

Starting slowly is not just for the novice, either. The map of a circuit may never change, but the amount of grip the surface offers does, and often. Different rubber levels may leave the line grippy, but slightly slippery off line. Then there are wet leaves in the autumn, and undried patches under the trees after a night's rainfall to think about, while a track that is not often used will be 'green' at the start of the day, which means it will not offer so much grip.

So, there's certainly a lot to take in on your opening laps; and booking a session with an instructor would definitely help.

Instruction

Early in the day is the ideal time to get yourself a bit of coaching if you can, and most of the better track day operators will have instructors on hand who will sit beside you for a stint of 20 minutes or so. It will probably cost you – though usually it's not too much, and at some days it's even free – and sometimes you'll have to book a session with an instructor in advance of the day, but it's worth it. In fact, if you're new to track driving, a session with an instructor will find you far more speed than any bolt-on, go-faster bit of kit ever could.

Actually, if you're new to track days then the level and the cost of instruction might be something worth looking into before you settle on the day you want to attend. Ask if the instructors

ABOVE It's worth booking a session with an instructor every time, even if you're an experienced track day driver, as there's always something new to learn when it comes to track driving.

rather than warm-up laps, so make sure you have a good look around. Not all operators will run familiarisation laps, though; some will run a couple of yellow flag laps instead, while others will just let you get on with it.

Either way, it's time to fly solo soon enough, and this is where the fun really begins. But it's also where you have to be just a little bit sensible. I know, you're champing at the bit by now, after all that form signing and tyre-pressure setting, but just remember that a track day is exactly that – a day – so there's no need to set the world alight in the first five minutes.

Most accidents on track days happen right at the start of the day, and they're usually caused by drivers being caught out by the unfamiliar track, cold tyres or brakes, or through sheer over-

are ARDS qualified too, which means they've been accepted by the Association of Racing Driver Schools to judge if a driver is up to standard to race. Usually you will find they are, especially if the track day is organised by a member of the ATDO.

Often a registered ARDS instructor will be a jobbing racing driver raking in a few extra pennies, and it's interesting to note that quite a few well-known professional racers have instructed at track days. You might even be lucky enough to get a well-known Touring Car or Formula 3 driver sitting beside you.

But if you do get someone you've never heard of, don't let it put you off. There's no reason why a clubbie Formula Ford racer, or someone who doesn't even race, can't make just as good an instructor as a hot-shot race-ace stopping off for

a giggle on his way to F1. In fact, some of the very best instructors at track days and at racing schools are some of the least successful racing drivers and, conversely, it seems fair to say that some of the best racing drivers can make poor instructors.

Maybe that's to do with temperament? After all, if there's one thing a good instructor needs it is patience – something that is not a major feature in the psychological make-up of many a wonder-child racer, I'd guess.

One problem I've noticed with a tiny, tiny minority of instructors who are also racers is that they do not seem to understand that you might not be keen on screaming every last rev out of your car or taking it to the limit on every corner. There was even one track day I attended where

ABOVE Ready for the off: a marshal will check armbands at the end of the pit lane before a car is allowed on track.

the only offs throughout the morning were when one particular instructor was sitting beside the drivers! That's very, very, *very* rare, though. So, don't let a bad apple put you off. And drive to a speed *you* feel comfortable with, not the pace that young Fernando Testosteroni, or whoever it may be, feels you ought to be going at.

Another quality instructors need by the bucket load is the ability to put people at their ease. Tensing up is a common problem with novice drivers, and the resulting stiffness of the arms and white-knuckle grip will lead to jerky movements of the steering wheel. A good instructor will sense if you're nervous and will do what he can to help you relax and enjoy your day.

Once in a while an instructor may even spot a rare talent, a 'natural' behind the wheel. It doesn't happen often but it does happen, I've been told, and then there's not a lot the instructor needs to say or do.

More usually an instructor will just sit beside you as you lap, pointing out the line, making comments, and giving you an idea of what gear you should be in. If you're in a noisy open car the instruction is carried out via hand signals and a thorough debrief once back in the pits – although some will have nifty little intercom systems which fit inside crash helmets and these tend to be pretty good.

Making use of an instructor is not just for

BELOW It's vital you pay regular attention to your mirrors during a track day.

novices, either, and many experienced circuit drivers will book a session on their first visit to a new track, or just to help polish their track craft. Some will even book coaches (some of whom are well-known motor racing figures) for an entire track day, as there's always something new to learn when it comes to the art of track driving – even those well-known coaches will admit that.

Sometimes, if you're happy with someone else driving your car, it might be worth letting the instructor drive it with you in the passenger seat. It can be a real eye-opener, showing you what's possible and what you should be aiming for. Most are amazed at just how much speed a top driver can take into a turn, and also the speed they're able to achieve with seemingly very little effort, as a good driver will always make the very difficult look very, very easy.

Also, because instructors are very often racing drivers, sometimes with huge experience of sorting cars, it's an ideal chance to get some feedback on how your car performs, and what steps you could take to improve it.

Yet, despite all the obvious advantages, many simply don't bother with instruction. That's their choice, of course, and there's something to be said for simply wanting to get on with it on your own, learning from your mistakes as you go. The only problem with that approach is that it can prove expensive in the long run.

What's more worrying is the ego thing. Many track day drivers, even beginners, just seem to think they have nothing to learn. Well, everyone has something to learn when it comes to track driving, and those who don't accept this probably have more to learn than most. Confidence in your driving is fine, but over-confidence can lead to mistakes and crashes.

The problem is that no one likes their driving to be criticised, do they? Many of us have had our driving licences since we were 17, so driving is almost as natural to us as walking, and as unashamed petrolheads it's what we're all about. And, as someone once said, to criticise a man's driving is tantamount to criticising his performance in bed.

But if I'm doing something wrong I'd like to know what it is, and so do many others. In fact, even professionals at the very top of their game look to coaching and are not above taking a little advice on technique. They may not talk about it often, but even quite a few Formula 1 drivers these days will rely on driver coaches to point out where they're going wrong. So, rest assured, investing a few quid in some instruction is certainly no slur on your ability behind the wheel.

Mirror, signal, manoeuvre

One thing an instructor will sometimes do (personally, I'm not sure of the wisdom of this) is take over the mirrors for you. The idea is that they concentrate on the rear view while you concentrate in the track ahead. But the problem with that is that you should be getting into the habit of using your mirrors as much as possible right from the very start of your track day career, so it's not something you want to take a break from. Either way, you certainly need to make it clear between the two of you who is actually taking responsibility for the mirrors.

This should be obvious really, but using your mirrors is absolutely vital on a track day, for while they're far safer than races, on the whole, they do tend to have one thing most modern races tend not to have, and that's wide disparities between the performance of the cars and between the skills of the drivers. Sometimes other cars can be upon you before you know it; they almost seem to pop out of a parallel universe, so you need to keep your wits about you. Make a point of glancing in the rear view as often as possible, and check the wing mirrors too. Knee-high specials and bike-engined prototypes have a habit of sneaking in under the radar, and the first thing you know about them is when they flash past at twice the speed you're managing in a blur of low-slung colour and a wail of revs – which leads us nicely on to the thorny old subject of on-track etiquette.

BELOW Good manners and proper conduct are exhibited by this Lotus driver, pulling over on the straight to let faster drivers through.

Etiquette and overtaking

At track days the onus is firmly on overtakers when it comes to getting past. It's up to them to make the move, but only where the move is allowed. However, that's not to say you should just happily lap with a long line of cars forming behind you, for that's a sure way to win the most unpopular driver at the circuit award.

So keep an eye on your mirrors, and when a faster car comes up behind you pull to one side of the straight (you'll be told at the briefing which side is for overtaking – it's normally the left in the UK) and let the faster car go by. Sometimes overtaking will be by invitation only – this will also be covered in the briefing – which means the car in front has to indicate it's moving aside before the other car overtakes. This seems to work pretty well, but you do have to be careful, as some specials will not have self-cancelling indicators and it's not unknown for drivers to lap with their indicator constantly blinking because they've forgotten to turn it off and have not heard it because of the snugness of their helmet and the noise of the engine.

But the important thing is to remember to let others past. It can be a bit unnerving when you first start out, and it might seem like you're constantly ducking out of the way of other cars, but try not to let it worry you. Also, while it's important you keep an eye on your mirrors, do not get too fixated on the cars behind, you've as much right to the track as the quicker cars, so while you should be aware of the cars behind you, don't let it spoil your enjoyment or ruin your concentration on the track ahead. It's a balance really, and one that some get very wrong – usually

very slow drivers in very fast and expensive cars, as you'll soon notice!

When you do get up to speed and it's you who's doing all the passing, then you need to remember that others might not be quite so confident; so take pains not to crowd people. It should be obvious whether a driver has seen you in his mirrors or not, and if in doubt, back off. There's always the next straight.

Almost all track day operators will stipulate that there's no overtaking in the corners or the braking areas for the corners. To put it another way, you can only overtake when the car is accelerating and going in a straight line. This means you can fully concentrate on your line if you're the car in front in a turn, but it also means that queues of cars can form when there's a series of corners. This can be particularly frustrating if you're in the car following, especially if you're quite a bit quicker, but it's something you just have to live with, until the next straight, at least.

It might even be worth hanging back before a corner so that you get a better run through it without the slower car baulking you. That way you get more momentum on to the straight so you're much quicker there, giving the other driver a very clear message that you're faster and should be let through.

One thing you should never do is flash your lights. It might be all right for hotshot Touring Car drivers carving their way through the back-markers but there's absolutely no place for it on a track day. It suggests aggression at the very least, and it will distract the driver in front at the worst – and you might even find yourself caught up in his moment as a result of it.

BELOW Overtaking is usually on the left and on a straight piece of track only.

As someone who, no doubt, sees track days as an escape from the grind of daily motoring, you'll be relieved to hear that traffic is seldom a problem on well-run days. The number of cars let out on the circuit at one time is always restricted, usually to something close to that allowed for racing, and they're let out at intervals. Incidentally, there's always at least one marshal at the end of the pit lane whose job it is to check the wristbands of drivers and passengers and then wave them on track.

Sometimes little knots of cars travelling at a similar speed do form – which in the main leads to everyone following the guy in front and picking up his bad habits. So watch out for that, and concentrate on your own driving, not the driver in front. However, if you do find yourself stuck in traffic you might consider backing off a little, giving your car a breather until you find a nice gap in the traffic – just like an F1 driver in qualifying.

If you're new to track days, it might all seem a bit frantic at the start of a day. The first hour is always busy, as everyone wants to get out on the track and purge those M25 blues. Don't let this put you off, though, as things will soon settle down. But if, to begin with, it does seem just a bit too crowded for comfort – and maybe a little crazy – then simply wait an hour or so. It will get much less intense as the day goes on.

Incidentally, if you have to slow drastically because of a mechanical problem, do your best to keep out of the way of other cars and try to let them know you've a problem, either by switching on the hazards, or by raising your arm – the racer's way – if you're in an open car. And if it stops out on track, then stay put, do not get

out of your car – unless it's on fire or is in a very dangerous position – and most definitely don't open the bonnet and attempt to fix it while other cars are still whizzing past.

Flag signals

If you're holding someone up for a while then there's a good chance you'll be shown a blue flag by one of the marshals. That said, flag marshalling differs from track day to track day, ranging from none to full use. Most will definitely make use of yellows, though, which warn of danger ahead and should be taken as a signal to slow down – though do not back off immediately and certainly check your mirrors before you do slow, as the guy behind might not have seen the flag. Yellow flags might also be supplemented by lights at some venues.

ABOVE Caught in the act! Overtaking on a corner is usually forbidden, and chances are that this Porsche driver was hauled in for a dressing down soon after this manoeuvre.

BELOW If you do get caught behind a slower car you need to be patient and wait until it lets you through, or make yourself a little space by slowing down and dropping back.

FLAG SIGNALS

Most track days will use a system of flags based on those used in racing, as outlined in the MSA Blue Book. The most important flags that could be used on a track day are:

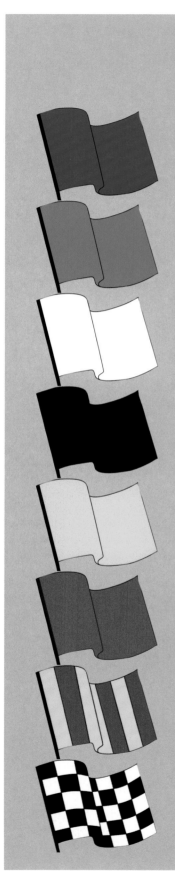

Blue. Overtaking flag: someone faster is trying to get past. Pull over on the straight and let them through.

Green. Track ahead is clear: go for it.

White. Slow moving vehicle ahead: watch out for a course car or the like.

Black. Naughty, naughty: you've done something the organisers aren't happy about. Time for a little chat. Either that or there's something amiss with your car. In both cases you need to pit.

Yellow. Danger: slow down – but check your mirrors first!

Red. Session's stopped: slow right down (check mirrors first) and pit.

Yellow/Red. Slippery surface: oil or fresh fall of rain – or other objects on the track at some track days.

Chequered. End of session: fun's over for now, so cool it – literally, slow down and cool those brakes.

And the same goes for reds, which means the session's been stopped and you need to slow right down and return to the pits, not stopping on the track unless you're signalled to do so. Another flag you might see (if you've been a bit naughty or there are bits falling off your car, or you're leaking oil, and so on) is the black flag. If this is waved at you then pit as soon as possible and go to race control – or wherever the organisers are – for a slapped wrist or an explanation of what was wrong with your car.

The red and yellow striped flag is often used, too, and this means slippery surface ahead, although it's also used to signify just about anything lying on the circuit on track days – including exhausts that have dropped off cars, or even a dead, and quite squashed, crow at one day I attended. Most flag signals will be based to a degree on those used in racing, so it's worth memorising what these are (see the panel).

End of session

One flag you're sure to see at some time during the day is the chequered one. This signifies the end of the session – or, if it's an open pit lane day, the lunch-break or close of the day. Incidentally, on open pit lane days the onus really is on you to come in at regular intervals, giving yourself and your car a rest. It's common sense really, because you should never run the risk of running out of petrol, pads, concentration and – as a result of the last – road. But when you do come in, make sure your intentions are clear, either by indicating, or putting your hand in the air if it's an open car with no indicators.

When you decide to take a break or when the flag waves for the end of the session, slow down and take your time around the last lap – treat this as a cool-down lap for the car. Use the brakes as little as possible to allow them to cool, but be sure to drive off line if other cars are still lapping at pace, then bring the car in and park up in the pits if you have a garage, or in the paddock if not. Be careful not to overshoot your pit garage, though, as on most tracks reversing in the pit lane is forbidden and you may be forced to do another lap before you can park up – though there's often a handy gap at the end of the pit lane which takes you to the rear of the garages. On the subject of the pit lane, do not treat it as part of the race track, and keep your speeds down.

Once parked, it's important you do not pull on the handbrake, or rest your foot on the brake pedal, as this could result in bits of hot pad material fusing to the discs, or even warped discs.

Running checks

It's worth checking tyre pressures and the tightness of your wheel studs throughout the day. Some suggest a quick check after the first session and then maybe a couple more. If you find that just one stud needs tightening, there's a fair chance it's stretching, so replace it. Also, make sure you check oil and water levels before you start the next session. You may want to keep an eye on tyre wear, too, particularly if you have to drive the car home – but more on all these points in Chapters 12 and 13.

There's another thing you might want to think about at the end of a track day, too. The final hour or so can be a bit less busy as some will leave, so it's often a great time to get quality track time with a little less in the way of traffic. But it's also the time – along with the very early laps – when accidents seem to happen, and it doesn't take a genius to figure out why. You're probably far more tired than you think by now. Concentration takes it out of you more than you would perhaps imagine, and almost certainly you've been up since stupid o'clock – an early start seems to be one thing track days do have in common with race days.

Chances are that adrenalin is getting you through now, so think about that, and if you realise you're not looking through the corner but focusing just on the track in front of you, then it's likely you're not concentrating properly. One Nürburgring regular once told me that when you start to read the graffiti on the track surface (which is a feature of the 'Ring), then it's time to call it a day. Good advice, that.

Personally, I always resist the temptation to do 'one last lap'. It might be just superstition but it seems to me that once you've decided to call it a day, then that's exactly what you should do. The last time I did 'one last lap' it was at a Brands Hatch day. I then found a little Peugeot 106 had spun at the bottom of Paddock Hill Bend just in front of me and was rolling slowly across the track. I missed it – just.

As well as making sure the car is topped up with fluids throughout the day, make sure you are too. It's easy to get dehydrated on a track day without really noticing it, so it's worth stowing a couple of litres of water along with your tools. As for food, then small amounts and regular is the obvious advice – that fry-up in the paddock diner might be tempting but you don't really want to feel sluggish when you're lapping at speed.

Once you've finished, let the car cool and then it's just a matter of driving home. Be careful here, though. Because you've been driving at high speeds all day, even regular motorway speeds will feel slow. So keep an eye on that speedometer, and watch out for the boys in blue – some of them have been known to wait outside circuits when they know a track day is on, speed guns in hand. Though at least this will remind you why you took up track days in the first place.

5

The driving platform

The driving platform

Not even the fastest of road drives will come close to what you'll experience at a race circuit. So you'll have to adjust your approach when you get behind the wheel at a track day. You'll also need to understand what the car is doing beneath you... and, for a start, you can forget some of the things you were taught to help you pass your driving test.

Ah, getting behind the wheel – the bit you've been waiting for. So, are we sitting comfortably? Actually, that's more than just a turn of phrase, for your seating position is *very* important. Your direct relationship with the steering wheel and seat will dictate the way you read the road – all those signals that are travelling up from the tyres and the road surface via the rim of the steering wheel, and the shift of the chassis' weight under your backside. So your ability to fully control your car will be greatly enhanced if you're sitting correctly and, as I said, comfortably.

It might seem like a basic point, but you'd be surprised how many track day drivers get it wrong. You'll see them with their arms stretched out straight, locked stiff, or with shoulders hunched over, arms jack-knifed and face close up to the wheel like they're about to take a bite out of it. Needless to say, both extremes are ill-advised – although it has to be admitted that the former never did the 1960s F1 aces much harm, but things were different back then, with light steering loads and an emphasis on small frontal areas to cut down drag.

But, as far as driving on a modern track day is concerned, you should be looking for somewhere in between the two extremes. Basically, you want your arms to have the room to be able to steer without becoming fully extended. So sit as upright as you can in the car and press your shoulders back against the seat back – remember, you want to 'feel' what the car is up to so you need a solid contact with the seat – then grip the top of the wheel with your fists. There should be a bend in your arms, and more of a bend when you grip the wheel at the correct (see below) position of quarter-to-three.

Another good way to set your seating position is to stretch your arms out in front of you over the top of the rim of the wheel. If your wrists rest atop the wheel, then that should be about right, but be ready to readjust for comfort should you feel the need. And make sure you can still turn the wheel through 180 degrees without your elbows snagging the top of your thighs.

You should also make sure that you can reach all the controls from your driving position, and that you don't have to stretch any of your limbs to get to them. Nothing should be fully extended. Your left leg should have a little bend still in it when the clutch is fully depressed, and the same should be true of your right leg and the other two pedals to be sure you can get all of the travel out of the throttle. Also check that you can comfortably select each and every gear and reach any important switches.

While we're on the subject of the controls, use the balls of your feet on the foot pedals, as these are not only the stronger part of the foot, but also the most sensitive. Ideally, if you've a footrest for the left foot, then use it, for not only will this help you to resist resting your foot on the clutch pedal, and perhaps risking riding the clutch, but it will also help you brace yourself in the corners.

It's also vital that the seating position you choose is as comfortable as you can make it, as this will help you avoid fatigue. OK, you might not be overly-worried about this, but tiredness can catch up with you and maybe even catch you out. You would be surprised how tired you can get when you're lapping fast for an entire day. And when you're tired it's just so easy to lose concentration.

Steering

As for the way you position your hands on the wheel, look at somewhere between quarter-to-three (mentioned above) and ten-to-two, the

FAR LEFT Arms bent and as upright a driving position as possible is the way to go when it comes to getting the right seating position, but you need to be comfortable, too.

BELOW Hands should be around the quarter-to-three position and it's *definitely* best not to feed the wheel through your hands while you're on track.

Make sure you can turn the wheel through its full lock and that you can reach all the important controls.

first of which is far more common, and in which case most then hook their thumbs lightly over the spokes of the wheel.

The important thing is that you should try not to move your hands from this part of the wheel all the time you're on the track, so you can forget all that stuff dear old George at the local driving school taught you about how you should feed the wheel through your hands.

Keeping your hands in a fixed position will give you far more precise control of the wheel, as there will be a direct correlation between the movement of your hands and the movement of the steered wheels. But, more importantly, you'll always know when the wheels are pointing straight ahead, which is vital when the car gets out of shape during a moment.

You can even go as far as crossing your arms at hairpins – a definite no-no with old George – but this is not always practical, so many will shift the position of their hands, and assuming this is a right-handed corner here, to something like twenty-five-to-one, making sure they get back to quarter-to-three once they're through the tight bit. On a left-hander, of course, it would be twenty-five-past-eleven. Alternatively, it might be better to keep one hand at three, or nine, depending on which way the corner went, so that you've

still retained that reference as to which way is straight ahead.

If you really cannot get out of the habit of feeding the wheel through your hands – and old habits painfully learned can be hard to drop – then you might want to stick a tab of brightly coloured tape at the top-dead-centre point of the wheel, so that in the heat of a moment you at least know where straight ahead is. Incidentally, you might have noticed that many aftermarket sports steering wheels have such a feature, usually in yellow, but always useful.

However, the best advice is *definitely* to keep the same grip on the wheel at all times, and it's worth getting used to this as soon as you can, as it's more accurate, gives better feedback and – unless you're a police driver – it's simply the correct way to go about the business of high-performance driving.

One time when you might feel the need to feed the wheel, though, is when you've got the car well out of shape and you want to get the front wheels back to the straight-ahead position as soon as possible. In this case some drivers will allow the self-centring in the steering to help them, which is usually a good deal faster than the driver's hands. But, rather than simply letting go of the wheel, a good driver will let it slip through his or her hands, pulsing at the rim of the wheel when necessary. It's a technique they use a great deal in drifting, but as far as track work is concerned it's very much a last resort.

Another important thing to bear in mind when you're gripping the wheel is exactly that, the 'gripping'. A white-knuckle grip is one of the most common faults when it comes to newcomers to track driving – which is understandable as it's a symptom of nervousness. But you really should try to relax, as this nervousness can be transferred to the car, and can even unsettle it in the corners.

Also, if there's any unwanted movement coming up through the steering – perhaps from the front wheels hitting a bump or kerb – and your grip is too tight, then you can find yourself fighting it as you feel it through the wheel, rather than allowing the car to take its natural course and riding out the bump.

If you're not relaxed at the wheel there's a tendency to raise the shoulders, so look out for this as a symptom, and then try to relax if you can. That's not always easy of course, and even top drivers will sometimes find they need to work hard to feel relaxed behind the wheel, especially when they're in a high-pressure race situation. As you get more into track days, however, and you realise it's about leisure and enjoyment, and there's no pressure to perform, then that fine line between being relaxed and fully concentrated should be easier to find.

It is also important that you are *smooth* – a word we will be returning to again and again – and progressive as you turn the wheel: don't jerk it. You need to give the front tyres the chance to respond as you turn in, so think smooth rather than sharp. Like the other controls on a car, the steering wheel is not a switch – something else we will be coming back to again and again – and you should always look to feed in lock progressively.

Watch fast driving in the movies and it's like alligator wrestling, all arms and elbows, but in real life fast driving you want to be gentle with the steering. And you also want to use it as little as possible. Surprising? Well, maybe, but the thing is the straighter you can keep those wheels, the faster your car will go. We will come back to why this is so in this and the following chapters, but first let's pause here for a moment.

It's important to realise that track driving is one of those things in life that can look incredibly simple from the outside, and especially simple on TV, and yet it is very, very difficult to do well – which is why those F1 lads are paid so much. In fact, it will take most drivers quite a few track days before they're really up to speed, and however much you read about it, whether here or elsewhere – and in Appendix 2 there's a list of some recommended reads that go into track driving in much more detail – it will be no substitute for actual track driving. I hope these pages will give you a general idea of what it's about, but you'll only begin to *learn* once you

LEFT Yellow tabs on the top of a steering wheel can help you find the straight-ahead position in the heat of a moment. (*VW Racing*)

ABOVE All that stops
this Elise from flying off
the track are four tiny
contact patches where
rubber meets road –
scary, huh?

get out on track. Oh, and as we've mentioned elsewhere in this book, taking some of the instruction that's usually on offer is pretty much a no-brainer.

One other thing: some of the things we'll talk about in the next few chapters will actually come to you quite naturally, especially if you've a bit of ability, as you can often *feel* the speed of the car through a turn. With this in mind there's no real need to get over-worried about the theory, but at the same time it's useful to have a basic understanding of what is happening to your car, so that when it reacts in a way you didn't expect you can think it through, and be ready for it next time.

Contact patches

Everything you do in the car, whether it's braking, accelerating or cornering, will be transmitted through each tyre to the road surface via the small patch of rubber on each tyre that is in contact with the track at any one time. This is often just about the size of the palm of your hand – and it's called the contact patch. It's not much, is it? Yet thankfully it's more than enough to do the job, just as long as you don't ask too much of your tyres.

So your tyre clings to the road through this contact patch, and there are three things that

contribute to just how well it does this. There's its coefficient of friction (which is to do with the tyre itself and the road surface), there's the size of the contact patch, and there's the weight that is acting on it – or the vertical load. Because we're talking about driving here and not car modifications, the first two of these are fixed, but the vertical load is all important, and we'll come back to that soon enough.

First, though, let's look at the way the tyre is able to grip the road. There are two things you're going to want from your tyres: traction, for braking and accelerating, and cornering force. The tyre is able to provide both because of its elasticity – it's not made of rubber for nothing – and this elasticity is talked about in terms of *slip*. We won't worry too much here about the longitudinal slip (that is the slip when the car is accelerating or braking on a straight piece of road), which is usually expressed as a percentage, but it is worth bearing in mind that there will always be a trade-off between the two. To simplify, you can't hope to have 100 per cent acceleration and 100 per cent cornering force. But 80 per cent cornering force and 20 per cent acceleration is more like it – and we'll come back to this later on in the book.

What we really need to know about now is cornering force. Put simply, to develop any cornering force the contact patch needs to be turned at an angle to the direction in which the car is actually going. *Gulp!* Honest, that's not as bonkers as it might first sound. As we've noted above, tyres are elastic, and because of this, when they're turned, they deform to grip the road surface. Think of it as the tread twisting to grab the road, if you like. The angle between the direction the wheel – which is rigid – is facing and the direction the contact patch is facing, and hence the true direction the car is going, is called the *slip angle*.

Slip angles

There seems to be some confusion when it comes to slip angles. Some seem to think it refers to the angle of a car to the road in a skid, others to a steering angle. But, as we've said, it is the angle of the twisted contact patch, and the car's true direction, in relation to the way the wheel itself is facing.

To generate a slip angle you need a side force on the tyre, such as a cornering force. Interestingly, you need a slip angle in the tyre to generate a cornering force in the first place, but you also need a cornering force to generate a slip angle – the words chicken and egg spring to mind.

Cornering force will go up as the slip angle rises, and that's a good thing as you can take more speed through the corner. But, as with all

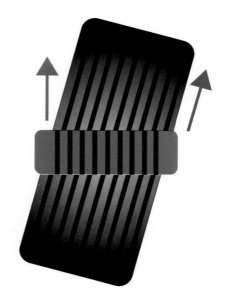

LEFT Slip angles. This is the angle between the true direction the car is going, which is in line with the way the contact patch is facing (red) as it distorts to grip the road, and the direction the wheel is pointed (blue).

good things, there's a limit, as a tyre can only be distorted so much before it cries enough.

You'll get most from a tyre at its optimum slip angle, and this tends to be in a range of 5° to 20°, depending on the tyres you're using. When you go beyond this the tyre will begin to lose its grip, quite slowly at first, but then quite quickly as the tyre springs back into its original state in a rubbery sulk.

This breaking away of a tyre, when it's at and then exceeds its optimum slip angle, is basically when a tyre goes into a slide, and the point before

BELOW Cornering force causes the slip angles to rise, and you get the best from your rubber at its optimum slip angle; after that it will break into a slide.

ABOVE As a car brakes into a corner its weight is thrown forwards – this is called dive.

BELOW As a car accelerates its weight is thrown back – this is called squat, as demonstrated by this beautiful BMW 3.0 CSL at a Santa Pod RWYB day. (*Rick Cuthbert*)

this is known as the limit of adhesion – it's the point where you can come unstuck! But the great thing is that it will give some warning before it does this, chiefly because you can *feel* the tread beginning to realign itself through the controls, seat, etc. Better still, when you overstep the mark, the tyre tread will not just let go, sending you flying into the gravel or worse, and this means you can exceed the slip angle and still drive the car in a slide, which then scrubs the speed off through the friction of the rubber against the track surface, which in turn brings the tyres back into their working slip angles. Cool, eh?

All this is *very* basic, and the way tyres work,

and slip angles in particular, is a subject that's worth a book on its own. Indeed, I thought long and hard before even mentioning slip angles, as they can cause more confusion than they're worth. That said, having a small idea of the way your rubber is working beneath you is useful, for it soon becomes obvious that not overloading your tyres is important if you want to make rapid progress through a corner and out of it.

Inertia

I really hate to resort to physics, but at this point I'm going to have to. Sorry. But let's try to keep it simple, shall we? When something is in motion it just doesn't want to stop or change direction. When it's not in motion it doesn't want to get going. This general laziness and lack of imagination on the part of matter is called inertia.

In a car we experience this all the time. Here's a commonplace example. We're approaching some traffic lights and we apply the brakes. There's a resistance to this that we can feel. The brakes are binding, slowing the car, yet the bulk of the car still wants to carry on going forward, and because of this the nose dips – this is called *dive*. Similarly, when we accelerate there's again a reluctance to get moving, and this time the rear of the car *squats*.

All this movement acts around the car's centre of gravity, which you might want to think of as the theoretical point at which you could perfectly balance a car, or more aptly, it's the fulcrum through which the body of the car will pivot. Because there's a force on the car at the level of the road surface – let's say we're braking here – this sets up leverage around the centre of gravity, which transfers weight to the front of the car.

Out on the race track it's the same. As a track day hero brakes for a corner, the weight transfers longitudinally along the car so that there's more of it pressing down on the front wheels, while the rear wheels are relatively light. The opposite occurs when he is accelerating out of the corner, the rear squatting as the front comes up, and now the front goes light. It does not take much to get the weight moving either, a simple lift off the throttle will do it – just try it in first gear in your road car in a quiet car park: accelerate, lift off the gas, and just watch that bonnet dip.

In a corner something similar happens, but this time we're considering a form of inertia commonly known as centrifugal force. Again, this acts around the car's centre of gravity, and it is a result of the cornering force set up by the tyres gripping the road surface, which is what is stopping the car from going straight on.

The higher the speed or the tighter the radius of the turn, the greater the centrifugal force, and the heavier the car the more centrifugal force the tyres have to resist, as there's more of that weight wanting to go straight on.

Weight transfer

So, we have weight moving along the car and across the car, though strictly speaking weight only acts vertically, but there's no point in confusing the issue here – you want to drive fast, not pass a physics exam. Earlier we mentioned that there were three things that are crucial to how much cornering force we can get from a tyre. These were the way it grips the road, how big the contact patch is, and the weight that's acting on it. The first two, as drivers on a particular day, we're pretty much stuck with, but the weight acting on a tyre is going to change. As we've just seen, this can increase because of the weight transfer across the car.

Now this might seem like a good thing, and it is – up to a point. But there's one thing that really spoils it, and that is this: there isn't a linear relationship between the vertical load placed on the tyre and the grip it can get. That means that while grip will increase as you apply more of a load on the tyre, the *work* the tyre has to do to get the grip increases at an even greater rate, so it needs to take on a larger slip angle, and we come to – and then exceed – that optimum slip angle

sooner than we would like. Bugger!

We can't stop weight transferring, but we must try to limit it if we can and, more important, control it. If a car is on the limit in a corner, any extra load on a tyre that is at its limit of adhesion can tip it over the edge. It's a delicate balance then, and any extra steering angle, cornering force, or weight transfer can upset it because it will bring about a greater slip angle. Yet, while we can't do too much about the amount of weight that is transferred, we

ABOVE As weight is transferred along and across this Clio, the front right bites into the track surface, while the rear left is off the ground.

LEFT Oversteer can be a symptom of weight coming off the rear or, as is the case with this lovely Vauxhall Chevette, lots of throttle in a rear-wheel-drive car. As a rule, people like oversteer.

can make sure it is transferred predictably, and in a way that will give us much more control over it – and the way to do this is to drive smoothly. It's all about *managing* your weight transfer, and very often it's also about making it work for you.

Oversteer and understeer

While a tyre will break into a slide when it's overloaded, there will always be an element of slide in the car itself, because of the yaw of the chassis as it turns – which is the *angle* of the car in relation to the corner, if you like – and just because grip will be unequal at either end of the car.

All this leads to understeer and oversteer, terms which you may well be familiar with as they've come into common usage over the past few years. If you're not, though, then the easiest way to think of it is like this: a car oversteers when the rear breaks into a slide, and understeers when it's the front tyres that are sliding.

Let's deal with oversteer first, simply because it's much more fun. This is when there's less grip at the rear than the front and the tail starts to slide while the nose will often turn in more as a result. It's as if it's steered too much into the turn, hence 'oversteer' (Fig. 1).

Oversteer can occur when you're turning into a corner or when you're in a corner, because weight comes off the rear of the car causing it to go light. Maybe you've braked into the corner, or maybe you've just come off the throttle – lift-off oversteer. You can actually use this sort of oversteer to your advantage, to quell understeer, particularly in front-wheel-drive cars.

To control oversteer of this kind you simply need to get the weight over the back end again by accelerating, which should shift weight back and give the rear wheels more grip, but you need to do this carefully and smoothly.

In rear-wheel-drive cars you can also get oversteer by treading on the gas too much as you're coming out of a corner, which is just asking a little bit too much of your tyres while they're still going about the important business of cornering. This is power oversteer, and it's controlled by simply easing off the throttle a little – though be careful not to do this too suddenly or by too much, as sudden deceleration will just bring more weight off the back and exacerbate the slide, maybe even turning it into a spin.

One thing most will know about oversteer is that you need to counteract the slide by applying opposite lock, but this is more of a natural reaction from the driver than an input, and the secret to getting your 'oppy locky' just right is simply looking in the direction you want to go rather than following the nose of the car – we will come back to this later.

Oversteer can be great fun, but understeer is usually just frustrating. That said it's the more common of the two at track days, mainly because modern car manufacturers always try to make a car understeer rather than oversteer, as it's thought to be the easiest situation to recover from if an unskilled driver gets into a bit of bother on the road.

Understeer is when the fronts have less grip than the rears and they're sliding across the road, meaning that the nose of the car will point away from where you're steering (Fig. 2). On track days this is often the result of accelerating too much, or too early – and not smoothly enough – in a corner. This causes weight to come off the front. But it

RIGHT Fig. 1. Oversteer.

FAR RIGHT Fig. 2. Understeer.

can also be the result of carrying too much speed into a corner.

The problem many beginners have with understeer is that they will wind on even more steering to try to make the corner. It's a natural reaction, but one which makes matters worse, and you need to fight the urge to wind on more lock. A tyre isn't designed to work when it's at a radical angle to the direction of travel, and they generally perform better when you aren't asking too much of them, so you really want to think about applying a wholly unnatural sort of opposite lock, pointing away from the inside of the turn to allow the front tyres to grip, and then getting on with your corner once the grip is retrieved.

You'll need to slow, too, to get weight back over the front, but you must not lift off the throttle suddenly. Just slacken off the accelerator a little and let the nose tuck back into line. It's very important in understeer – and oversteer – situations not to think of the accelerator pedal as an on-off switch. You'll usually find that you can lift the accelerator just a little bit to get the car to turn in nicely, and then the weight transfer along the car will not be too violent – otherwise it might force you into snap oversteer and perhaps a spin.

Sometimes, with powerful front-wheel-drive cars, you might find the car understeering through a turn because the front wheels are spinning, a sort of 'power understeer'. This is a common problem in the wet. The above also applies here

but, again, you have to go easy when you're lifting off that throttle.

There's also a neutral state which, at the limit, can result in the fabled four-wheel-drift, where the car is perfectly balanced and all four wheels are sliding. It's difficult to achieve in modern cars with modern tyres, but if you are four-wheel-drifting, then it means that you're driving like a champ and your car is beautifully balanced. Incidentally, you'll often find this is a useful state in fast corners, where you might be more interested in carrying speed than maintaining traction.

A driver's platform

I like to think of all the above as a driver's platform. We're sitting comfortably, and as we drive we feel – through the steering wheel, the seat, and the controls – what the car and the tyres are up to. We know that we need to work the tyres to get the best from them (to generate an optimum slip angle), and we know that we should try not to overload them. Because we understand that weight moves across the car, forward and back and laterally, we can use this to our advantage – maybe squeezing the throttle to move weight to the back for more grip when it's oversteering, or easing off the throttle to move weight to the front to dial out understeer.

So we have our platform – the stage on which we are to perform – now we need to learn our lines…

ABOVE Note how this car seems to be going straight on even though there's still quite a steering angle wound on. The front tyres are actually sliding across the track here and the car is understeering.

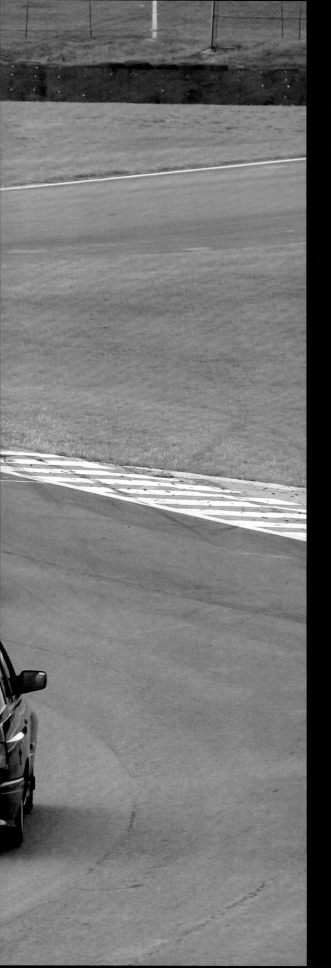

Learning your lines

Learning your lines

It might be invisible, but the racing line is the most-talked-about feature of any race circuit, and finding, then keeping to, the right line is the key to getting the very best from your car and your driving at a track day.

In a way, a car driven well on track is a bit like a slot car. It follows a groove, but an almost invisible groove, and it strays from this as little as possible. This groove is called the racing line, and it's quite simply the fastest route for a particular car around the circuit.

'But hang on a minute,' you're thinking, 'if a track day is non-competitive, why would I have to worry about anything which calls itself the 'racing' line?' Good point, but the fact is that by keeping to the line as much as possible you'll get far more enjoyment from your track day. The car will feel faster, and it will *be* faster, so there's less chance of you having to pull over to let others through, which can be frustrating if you're doing it all day.

More importantly, there's also the fact that the circuit tends to give less grip offline, as anyone who watches Formula 1 on TV will know. Little pieces of debris and worn-away rubber will be swept from the racing line by the constant passage of the cars through a turn, and these will settle off the line – meaning that those areas can have a lot less grip. Obviously best avoided then.

Classic line

If it wasn't for those pesky laws of physics – damn them and their meddling ways – the quickest way around a circuit would obviously be to hug the inside of the track as a middle distance runner would. The problem with this approach in a car is that you would be tightening the turns, with the result that you would have to slow down to make them. But by taking the widest radius possible through a turn, even though we are actually spending more time in the corner, we are able to carry much more speed through it, and this more than makes up for any ground gained by short-cutting around the inside.

This is the classic, sometimes called the 'geometric', line through a corner, and it describes as shallow an arc with as wide a radius as possible. This arc should almost touch the edge of the track at three points, the turn-in, the apex, and the exit (Fig. 3).

Because you want to make the arc as shallow as possible for fast lapping, it's important to get as close as you can to these three points. Indeed, racing drivers are usually centimetres from them: centimetres from the outside at turn-in and exit,

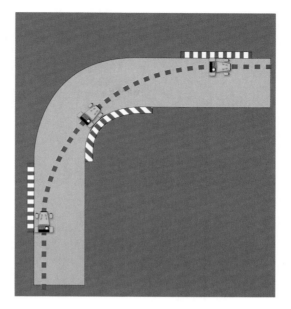

FAR LEFT It makes sense to follow the racing line as much as possible, as not only will it mean you're faster but it will also help you avoid all the rubbish that is swept off line.

LEFT Fig. 3. The classic or 'geometric' line through a corner describes a smooth arc through the entry, apex and exit.

and centimetres from the inside at the apex. They're 'using all the road'.

But let's pause here. At a track day you don't need to squeeze every last tenth out of a lap, so maybe it's worth giving yourself a little more space to the edge of the road than a racing driver would, especially under braking for the turn-in. That way, if you do snatch a wheel when you hit the anchors, you'll have room to do something about it, and you won't veer straight on to the grass which could pitch you into a spin.

Incidentally, this 'using all the road' is one of

LEFT This Civic Type R's using all the road as it comes out of the corner.

RIGHT Fig. 4. The classic line will still be used on fast corners where there's little acceleration at the exit.

FAR RIGHT Fig. 5. Note how you keep the speed gained from a quick exit all the way up the following straight.

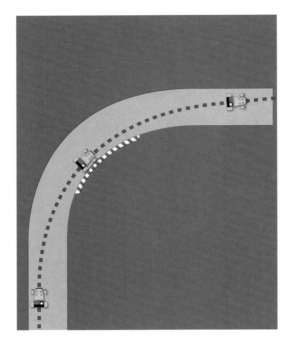

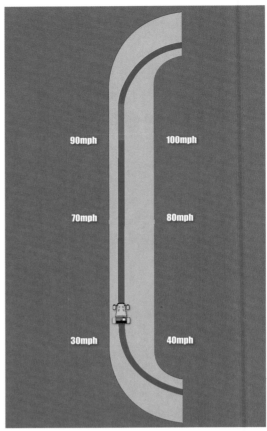

the most obvious differences between driving on the road and a track day. Use all the road on the highway and you're likely to shorten your wheelbase substantially – against the truck that's coming the other way...

But back to the track, and the classic line through a corner, which goes something like this. You arrive at the entry point to the corner (the turn-in), steer through the apex, which means cutting across the entire width of the track in as shallow an arc as possible, kissing the apex with the inside wheels at the point around which you're turning, before continuing this arc out to the exit. Sounds simple enough, doesn't it? But there's a bit more to it than that.

Slow in, fast out

While the classic line through a corner is also the quickest way through a corner, you need to bear in mind that the thing about track days is that they take place on tracks. And the thing about tracks – even the twistiest of them – is that they're made up mostly of straights; or at least stretches where you're on full throttle (which, to keep things simple, we will refer to as straights as it amounts to the same thing).

Obviously, then, to go fast you need to maximise your speed along the larger portion of the track, which is down the straights; which in turn means maximising your speed *out* of the corners. This is because the faster you accelerate on to a straight the greater speed you'll ultimately gain down that straight – right to the very end of it, in fact.

It's worth stressing here, though, that a classic line will still be used on wide and fast corners (Fig. 4) where there's no acceleration at the exit anyway, and that carrying the speed through these turns can be one of the great challenges of track driving.

But back to maximising speed on the straights. Think of it this way: you hit the straight coming out of the corner at 30mph then keep on accelerating up the straight until you hit the braking zone for the next corner at 90mph. But, if you come out of the corner quicker, say 40mph, then you'll come to the same braking zone at 100mph. OK, you might have to brake a little earlier, but the time lost here is marginal when compared to that gained all the way up the straight, because braking areas are very short when compared to the portion of the straight you'll be accelerating along.

Also, at each stage of the straight you'll be quicker. So, if on exiting at 30mph you'll be: 40, 50, 60, 70mph and so on at successive points up the straight, exit at 40mph and it will be 50, 60, 70, 80mph at those very same points (Fig. 5). When you add that up across a lap, that's a lot of the lap you'll be quicker on.

With this in mind, what you try to do in a corner is to feed in the power as soon as possible. But

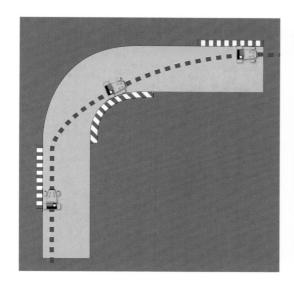

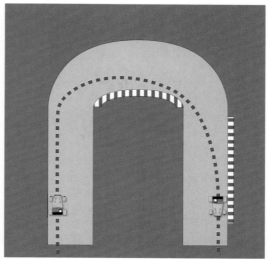

because the tyres are at the very limit of adhesion through hard cornering, any added acceleration will cause them to break away into a slide. So to balance this out you need to straighten the steering wheel towards the exit of the corner, trading off cornering force for acceleration (which we will go into in greater detail in the next chapter). By doing this you'll also increase the radius of the line's arc at the end of the corner.

But if you do this on the regular classic line you'll just run out of road, so you need to modify the first part of the turn so you can straighten the car out earlier, usually by turning in later and

therefore sacrificing some speed in the first part of the corner (Fig. 6).

Study the way a driver takes a hairpin (Fig. 7) for a good example of how it works. The first part of a turn is sacrificed by him letting the car go in deep, then turning-in tighter in order to be on the throttle sooner, meaning that the car is pointing in the right direction with the tyres at a lesser angle, with less cornering force, much sooner.

This often confuses people who are new to the track, those who have arrived with the assumption that track driving is all about late braking and driving a corner as fast as possible. It is not. It

is about driving a *lap* as fast as possible. This technique is called 'slow in, fast out', but I like to call it SIFO, because you hear it so often it deserves an acronym.

Pairing corners

This sort of thinking should also be applied to a series of bends. For instance, if a pair of linked corners, like an S-bend, were followed by a straight, then the speed through the first of the corners should be sacrificed in order to carry more speed through the second and ultimately on to the straight. There's no point in rushing through the first then finding you've compromised the line for the second; it's all about the speed you take on to the following straight, remember.

Watch some race drivers through a chicane for a good example of how this works, and note how they will surrender their ideal line through the first part by brushing the first kerb very late, which will then give them a good run at the second apex and the space to wind the car up for the straight beyond it (Fig. 8).

However, sometimes it can be worth going into the first of a series of corners faster than you would if it was followed by a straight, but that's only if it's at the end of a long straight and there's nothing to be gained from a quick exit out of the second corner, maybe because it's the start of a fiddly complex of tight corners. And even then the speed gain will certainly not be as marked as a SIFO line through a corner leading on to a straight.

Sometimes you might need to treat two corners that flow into each other as one, maybe treating it

as a double apex turn (Fig. 9) or perhaps simply ignoring the first apex and concentrating on the second. There are many other variations on the theme, all of which you'll arrive at by remembering that you want to maximise your speed up a following straight.

Variety's spice

There are plenty of things which will muddy the water when it comes to lines, and not all corners will fit a simple turn-in/apex/exit pattern. For instance, some long corners might have more than one apex, or even one very long apex, because you need to keep the car tight to the inside of the track before letting it drift out to the exit.

Some cars will even take different lines from others, and this is especially true of powerful rear-wheel-drive cars at tight corners, where the aim is to get on that power even earlier to play to the car's strengths. This is certainly worth bearing in mind, but it's probably still best keeping to the regular line on a track day at first – if it's marked with bollards (see below) – and then experimenting with lines that suit your car best as you get more experienced at track driving.

But the vital thing to remember is to treat the track as a whole, and think about the consequences of the line you choose. And remember, even though you're not racing, following the racing line will give you more grip and speed, and it will also mean your route for a turn will be predictable by the other drivers. So, follow the line, and you'll do just fine.

BELOW LEFT Fig. 8. In this chicane our car has taken a late apex at the first part to get a better run at the second part and hence a quicker exit on to the straight that follows.

BELOW MIDDLE Fig. 9. Sometimes it can be better to treat two corners as one, making it a single corner with two apexes – a double apex turn.

BELOW RIGHT Fig. 10. It's difficult to get the scale right for this, but just pretend that car's a bit smaller, in which case the apex can be stretched right around the bend.

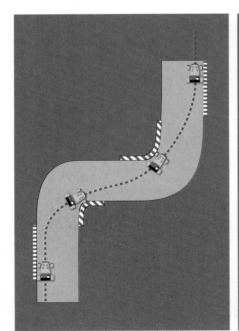
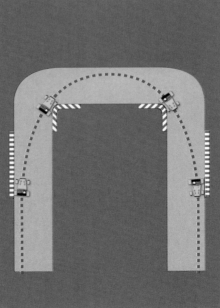
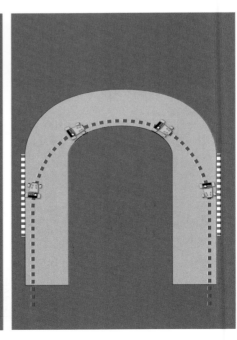

Homework

These days there are a number of ways you can
sneak a peek at a circuit before you actually get
there, and one of the real breakthroughs has to be
the development of onboard camera technology over
the last couple of decades. Back in the early '80s,
onboard cameras were heavy and unwieldy devices,
and there wasn't so much in-car footage about, but
now they're tiny things and a televised race would
seem incomplete without a bit of in-car action.

Not all circuits will have races covered on TV,
of course, but many track day and club race
drivers will take a camera onboard and record their
efforts. Often these will find themselves on the
internet, so it's always worth doing a little Googling
before you set off for your track day.

You have to be a bit careful, though, for the
line some hotshot takes in a big power GT, or
the line a car with an aero package on it will take,
might differ from that you'll be using in your little
Pug 205. Also, don't assume that the guy you're
watching is doing it right; there are plenty of clips
of bad driving on one very famous access-to-all

internet site, so be careful.

Another thing about film/video is that it rarely
does gradient well, and this is something you
should bear in mind. You won't fully appreciate just
how steep Paddock Hill Bend at Brands Hatch is
until you've driven through it – although 'over it'
might be a better way of putting this. Still, onboard
footage will at least help you find your way around
the circuit and, provided it's up to date, it will show
the various landmarks to assist you in formulating
your lines on the day.

The same goes for video games, with the
added advantage that you're actually using your
hands here, building up what athletes refer to as
'muscle memory', so that your brain has a mental
image of what is expected of it and what reactions
to take. But, again, games can only go so far,
and you certainly shouldn't rely on them too much
– just because you're flat out through Donington
Park's Craners on a Playstation doesn't mean you
can do it in real life. Might seem obvious, that, but
I've seen someone come a cropper because they
thought it did.

ABOVE Note the wide
entry for this turn at
Folembray.

instruction of a circuit guide with the immediacy of onboard footage, which has been done with the excellent *How to Drive* DVDs – available for Oulton Park, Brands Hatch, Snetterton, and Cadwell Park. Using a mix of in-car camera and graphics, racer and writer Mark Hales shows the line and explains how to drive it, what to watch out for, and which way round is both quick and safe. Highly recommended.

There's one thing you should always bear in mind when it comes to doing your homework before a track day, though, and that is that even if you do feel you know the circuit after using any of the above aids (you don't!), remember that all of these can become out of date. There's nothing quite as sobering as heading into that flat-out right-hander to find it's morphed into a second-gear chicane. The advice, as we've said before, is to start off slow and you'll be less likely to come across any nasty surprises.

Incidentally, as mentioned above, more and more track drivers are now using in-car cameras to record their laps. Thanks to technology they're remarkably small these days and the picture and sound quality tends to be excellent, while the price is not as steep as you might think. They're also a great way to improve your driving technique from the comfort of your favourite armchair with a cold one in your hand, picking up on small mistakes you might not have noticed in the heat of fast lapping. Most operators are happy enough for you to run with a camera in your car as long as it's fitted properly – but you should check with them first. However, passengers using hand-held cameras and phone-cameras are definitely a no-no out on track. I've even been at a track day where a driver has been black-flagged for trying to film *his own* efforts using a hand-held camera! It might have been funny, if I hadn't been following him at the time...

ABOVE You won't get a true feel for just how steep Paddock Hill at Brands Hatch is until you drive it.

BELOW Bollards are often used to show the turn-in, apex (as shown), and exit points of a corner.

Perhaps the best way to get a preview of a circuit – I'm resisting the temptation to say 'learn' here as that can only really come on the day – is to invest in a circuit guide. In the UK there's been a good guide for all of the circuits for some time now in the shape of the *UK Circuit Guide*, while the same company also publishes guides for the main European venues, plus one for the Nürburgring. The guides break down the circuits into annotated corners and then take you through the line, pointing out areas of danger and where to find time. They also come with all sorts of other useful nuggets of information, such as places to stay and where to get fuel.

But maybe the best way to get some prior information of a track is to link the detail and

Talking bollards

Most track day organisers will mark out the line with bollards. Often, though not always, these will be in a traffic-light order. So, you'll find a red cone at the turn-in, a yellow at the apex, and a green at the exit (Fig. 11). At other times there will just be cones at turn-in and apex, while on some track days there are no bollards whatsoever. Sometimes there will also be a cone, or even a board saying 'brake', before the turn-in cone to mark out the braking area, but as all cars have vastly different braking capabilities this is usually a rough guide, as well as being a useful way of showing the no-overtaking zone that generally lies within the braking area.

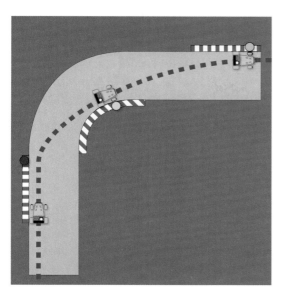

A word on airfield lines here. The outlines of these circuits are generally marked out with cones anyway, so bigger cones will be used to point out the corner reference points of turn-in, apex, and exit, though often they will not be marked at all. At airfields there will be a lot more room, too, and maybe a wide diversity in the lines being used, with plenty of scope for experimentation.

For instance, you'll sometimes find that a corner will lead on to a massive area of asphalt with no clear exit point. This means you can take a very early apex which allows you to exploit all the room at the exit. But, you also need to bear in mind what's coming next, and there's no point in going so far past the corner that you've made yourself a new straight. In these situations it's all about thinking ahead to the next turn, and it's something that adds to the appeal of airfield venues.

FAR LEFT Fig. 11. Sometimes a traffic-light system is used, with red bollards at turn-in, yellow at apex, and green at exit.

BELOW Sometimes braking areas are shown with boards, but these are usually just rough guides and as much to do with marking out the no-overtaking zone as showing you where to brake.

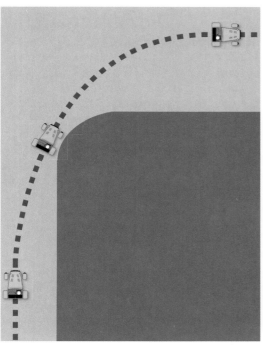

Looking ahead

The bollards are hugely useful aids when it comes to finding your way around in the early laps, but it's very important not to get fixated on the one closest to you at any given time. This is because one of the great secrets of driving a track fast is to develop the ability to look ahead through the corner. This means that as you're coming to the turn-in point you should be looking *through* the corner to what lies ahead.

The reason for this is to allow the car to flow through the turn. You should be trying to connect the cones by describing an arc past them, not by joining them in a series of little straights like the

ABOVE LEFT AND LEFT There can be a great deal of room at the exit of corners at airfields, which allows you to take an early apex to exploit the space, but then you also need to decide when it's right to bring the car back in order to line it up for the next corner.

edge of a 50p coin. So, as you arrive at the turn-in point you should be looking *through* the apex to the exit, focusing on where you want the car to *finish* up, rather than where you want to go to in the very next part of the turn. You also need to visualise the line you're going to take through the three points of the turn, as it's no use just driving to the exit without passing through the apex, so try to *see* your line through the turn in your mind's eye before you drive it if you can.

All of this is important because if your eyes are fixed on just the cone you're negotiating – or the apex if it's not marked – then you'll find yourself simply reacting to the next cone once you're past, upsetting the car's balance, rather than allowing the car to continue its natural arc towards the exit. You'll be dividing the corner up into a series of little reaction manoeuvres rather than the long flowing arc you're after.

Sometimes the exit will be hidden, for instance at corners where there's a blind brow, and then you need to visualise it. So make a mental note of it on your slow laps and paint a mid-air exit in your mind's eye, using the track furniture, trees, or whatever as points of reference.

It's all about seeing the corner as a whole, really, which can be confusing when everyone is going on about the *three* parts of a corner (turn-in, apex, and exit). But learn to see the big picture and you'll be well on your way to becoming a very accomplished track driver.

Finding the line

When there are no cones it's all a little bit more difficult – but, just like actors, track drivers have to learn their lines. And, also like actors, some learn them quicker than others. The real aces seem to have a sixth sense when it comes to choosing a line around a track and can be in the groove within a couple of laps. It's amazing how some drivers can do this. I've often supposed it was something innate, yet a natural understanding of something as unnatural as driving a car through a series of corners seems an unlikely ability to be born with, so chances are it's developed through experience.

That said, some people do seem to have a natural bent to play games like snooker and pool, where they just seem to have an inbuilt understanding of the angles involved, so maybe it's something similar? Then again, I once asked a very good race driver I know if he was any good at pool and he said he was rubbish! So maybe not.

If you're new to a track and there are no bollards to help you, then taking an instructor onboard for a session or a few laps is an obvious move. Other than that it's down to feel – you feel for how the tyres are performing beneath you – and a little bit of experimentation. There are clues, though, for although in the opening to this chapter we said the line was invisible, in many cases this is not strictly true – you only need to look at the tyre marks on the kerbing at apex and exit to see that. And these clues can be used to help you find the ideal line.

It might also be worth taking as late an apex as possible on the corners to begin with, this will at least mean you should have more room to play with at the exit. You then gradually move the apex back towards the first part of the turn as you find your way.

You need to think about gradients and cambers, too. Bear in mind that your car will brake and turn in, better on an uphill section of track than a downhill section, and try to make the time you spend on a negative camber – that is the road surface is sloping away from the inside of the turn – as short as possible.

BELOW It's important to look *through* a corner or a sequence of corners and not to get fixated on the bollard that's closest to you.

And how do you know when you've got it just right? You'll feel it, is the easy answer to that. The car will come out of corners quickly and smoothly and you won't feel that you have to fight it through the last part of the turn. Then there's the rev counter, or even the speedometer. If you're carrying more revs (or speed) out of a corner and on to a straight, then chances are your line is better than the lap before.

You should be able to tell if you've got it wrong, too. If you need to ease off the accelerator to stay on track, then you've obviously hit the apex too early, while if you've loads of room left on the outside at the exit, then maybe you've hit the apex too late.

Smooth operator

If there's one word those in the business of teaching people how to drive fast use the most it's 'smooth'. This is mainly because a smooth line is crucial if you want to *manage* the way the weight transfers across the car, and control how this acts on the tyres. This will pay dividends when it comes to outright speed, but for the track day driver – who hasn't got the stop watch to worry about – a smooth approach will also put less strain on the car. Smooth pays then, there's no doubt about that.

But while you need to be smooth with everything you do, the idea is to be quick as well, and that's where the art is with this track driving lark. Indeed, many novices do seem to confuse smooth for slow. You see it all the time; they've been told to keep it clean by an instructor, but in attempting to do so they're just keeping it slow. They're braking far too early and only accelerating as they leave the corner – and if it's speed they're after, then that's simply not what it's all about.

Having said that, though, if this is the approach you feel more comfortable with, fine, we're talking about track days here after all, and in the final analysis it's up to you to decide how you want to drive on a day. As we've said before, track days are all about freedom.

Kerb your enthusiasm

Racing drivers will always try to steal a bit of extra track by taking to the kerbs, because when every tenth of a second matters, every millimetre of the track matters too. But on a track day you're under no pressure to find that extra tenth, so it makes good sense not to rely on running the kerbs too much.

BELOW You can cut corners at race circuits by using the kerbs – like this MX5 at Llandow – but on a track day you have to ask yourself if it's worth risking damage to your car for a tenth of a second.

Why? Well kerbs can be rough – mainly because circuit owners like to discourage drivers from straying on to their nicely manicured grass – and they have even been known to damage cars. Hit them wrong and the shock can jolt right through the chassis. I've even shattered a piece of fascia plastic by running over one kerb a bit too violently. Also, kerbs are usually painted and can be slippery as a result, so be particularly careful about running them on the exit of quick corners.

That said, you'll find that some kerbs will be OK to use, and as they will often open out the corner they might be well worth running over. Indeed, sometimes kerbs will be flat, or not much more than *thruppy* little rumble strips to mark the edge of the track, and these should not upset your car too much. Try them at slower speeds first, is perhaps the best advice.

As for those kerbs at the inside of a corner – at a clipping point, for instance – then, in dry conditions at least, as long as they're not too steep they present less of a problem, because the inside wheels should be unloaded and doing little of the work and they should simply brush lightly over the kerb.

Chicanes can have very nasty kerbs, so bear that in mind before you straight-line them, especially on the way in to them when the weight hasn't had the chance to shift over to the outside of the car.

And remember, you're not in a race, and there's little point breaking a wheel or knocking your tracking out of alignment for the sake of a tenth of a second that means absolutely nothing. Oh, and one other thing about the kerbs, just keep well away from them when it's wet – but more on that in Chapter 9.

BELOW AND RIGHT
Think carefully before you use the kerbs at high speed. The flat ones are usually OK to run over, but they can be slippery.

Speeding up and slowing down

Speeding up and slowing down

We're on the right line then, but there's a bit more to fast lapping than that. You also need to understand how to use the brakes and the throttle correctly, and then there's the small matter of how you should change gear.

Many think there's something heroic about braking as late as possible. Indeed, the 'last of the late brakers' is a romantic motor racing cliché, conjuring up images of cars arriving at hairpins, smoke pouring from a locked front wheel. Looks good, but even in racing it's usually counter-productive and is often seen as a sign of desperation or over-driving – and a track day is not the place for either.

One track day operator has gone so far as to say that drivers trying to brake late cause more incidents than anything else on track days. Late braking is also a sure way to help cook your pads (see Chapter 12). But, interestingly, considering it's a habit most borrow from racing, it can also be a sure way to ruin your speed around a lap.

Let's take Donington Park's short circuit chicane

as an example here. After the chicane there's a straight stretching for about 700 metres past the pits and leading into the first corner. So you'll be looking to be accelerating down most of this straight. Meanwhile, the braking area, even though it is at the end of a far longer straight, might only account for about 50 metres or so, depending on your car.

Clearly then, the amount of time you can make in that 50 metres is negligible when compared to what you could lose through accelerating poorly into the following 700 metres because you messed up the corner trying to find time in the braking. So you might gain a tenth in the braking, but then lose half a second down the following straight.

But, much more important than that, by braking at the last minute you give yourself far less time to get out of trouble if something should go wrong, or if the brakes should fade, and you're also far more likely to spin because you've carried too much speed into the corner.

That said, you should try not to fall into the trap of braking too early and by too much, with the result that you find yourself trundling up to the turn-in off the brakes. You need to brake right up to the turn-in point so that you have that weight over the front wheels when you turn. But, to begin with, it's obviously better to brake too early than too late, then build up from there.

There's a great deal of talk at track days about braking points, such as the 'I'm not even thinking about braking before the marshal's post', countered by the 'That's nothing; I don't touch the brake before the oil stain on the kerb…' Ignore it. Not just for the obvious reasons of being sucked into an accident through trying to compete, either.

You see, the really important thing is not where you start your braking but where you *finish* your braking. OK, a general idea of where you should start is useful, but you should really try to focus ahead to the point where you're going to come off the brake pedal – the turn-in point. This is a much better reference, and you should judge the braking with the aim that it finishes at this point, rather than wait until you're passing a landmark and then hitting the anchors.

This is because every lap will be a little different, and as the day goes on chances are you'll improve and hit the preceeding straight quicker, thereby taking more speed into the corner, which means any earlier reference point will be redundant. And let's not forget the changes in the performance of the brakes over the day.

So look ahead to the turn-in point and judge your braking in relation to this and to how the car is performing at that moment, and don't get too hung up on braking points. Use them, by all means, and treat them as a rough guide if you do – oh, and use your own, not those that the bloke in the paddock café suggested.

On the subject of reference points, it's worth having a few of these over a lap to help you with positioning the car, maybe a seam in the asphalt which you know is half a width away from where you want to be in that part of the turn, or a building which is in line with the exit, something you can look ahead to, as discussed in the previous chapter. One obvious thing, though: photographers and spectators do not make for great reference points, simply because they're likely to go for a cup of tea.

Straight line braking

Most track day instructors will tell you to brake in a straight line, at least to begin with. You're already asking plenty from the tyres with the braking as it is, the last thing you want to do now is to ask them to turn as well.

With that in mind, when you start out on track days try to think of braking as a completely separate stage in the process of taking a corner, though one that goes on right up to the turn-in point. Later on you'll find that your skills will develop to the point that you might prefer the braking to bleed into the turn-in phase of the corner – but more on that later.

FAR LEFT It's easy to lock a front wheel if you're braking late or, as in this case, you're turning in while you're on the brakes.

BELOW It's important to be progressive with the brake pedal, both when you're applying the brakes and when you're coming off them.

And think *smooth*. It didn't take us too long to get back to that particular word again, did it? But that's the thing, apply it to all your driving inputs and you're halfway there, and if you can be smooth and fast with it then you might just have something special.

When drivers talk about smoothness in terms of braking, the word they also use is *progressive*. You really need to treat your brake pedal as if it's a living thing. Startle it and it will bite. So squeeze the brake pedal rather than jump on it. The car will respond much better to smooth movements on the pedal, but that's not to say your movements should be slow. With practice you'll

be able to brake progressively and at speed, and that's the trick.

Under braking the weight of the car shifts forward so that you have it pushing down on the front tyres, but you really want to ease that weight on to the fronts when you start your braking. Make the weight transfer work for you. If you stamp on the anchors all of a sudden the front wheels will simply lock up, but load some weight on to the fronts first then they will have more grip, so when it comes to the harder braking they're less likely to lock. Clever, eh?

But it's not just about the way you apply the brakes, it's about the way you come off them, too. It's no good just letting go of the pedal; you need to ease off the brake. Think about it. As you come to the turn-in point, lift off the brake suddenly and the weight will come off the front tyres violently. Ease off the pedal, though, and you'll have far better control over the way the weight comes off the front. In fact, the ability to come off the brake progressively, and yet still quickly, is one of the defining skills of the great racing drivers, and it's well worth practising.

Later on we will talk about how it can be an advantage to brake into the turn, but if you only ever brake in a straight line you won't be doing anything wrong. Meanwhile, you'll be having as much fun and thrills as the next man, and that's the most important thing.

The loud pedal

While you need to be sensitive with your middle pedal, the same also goes for your right pedal, too, and once again it's as important when you're taking your foot off it as it is when you're putting your foot down. Remember, any sudden deceleration can cause an equally sudden weight transfer, and sudden shifts of weight across the car are what you're trying to avoid. So you should always aim to *ease* off the throttle rather than just let it go.

Many beginners have trouble with their throttle control. You see it all too often out on track – those drivers who are treating the throttle as if it's an on-off switch. It's not. There's a wide selection of accelerator positions between full-throttle and no-throttle; remember to use them all.

This is just as important when you start to pile on the revs to come out of the turn as it is when you're lifting off. Resist the temptation to stamp on the gas as soon as you see the exit. You should be unwinding your lock and accelerating smoothly to match the steering, and getting ready to snatch the next gear on the exit or part way down the straight.

BELOW It's more important to be accurate than fast with your gearchanges.

Changing up

Now, I say 'snatch' above, but to be honest there's no real need to rush your gearchanges on a track day. It's also actually the same in racing, where accurate changes are far more important than lightning-fast shifts – in fact, the time gained in five laps of Billy Whizz shifting can be lost in one missed gear.

A missed gear is nothing more than a frustration on a track day, and in these days of rev limiters it will not even buzz your engine. It's an annoyance, though, and very embarrassing if you do it as you're driving past the pits!

Yet it's very easy to get caught up in the whole race driver thing and find you're going for the hand-blurring shifts. It's something about being on a track, I guess, and you can get carried away. But at least try to be accurate, and concentrate. It's so easy to come out of a corner and think the job is done. It's not; there are still clean shifts to make all the way up the following straight.

It's worth thinking about your gearbox, too. These can be expensive to fix, so you should be doing all you can to look after it. The main thing to remember is to be gentle with your gear shifts. So go easy on that stick, don't hold it like you're trying to crush it. On the up-change from second to third on a traditional H-pattern shift, for instance, gently nudge it into place with the heel of your hand. While from first to second and third to fourth cup your hand over the gear-knob, without gripping it too tightly, and smoothly ease it into gear. One tip I've heard is to try to allow the springs to do the work rather than forcing it in yourself, and this certainly seems to work with most cars – particularly when it comes to cutting down on that infuriating but common problem of shifting from second to fifth, rather than second to third as intended.

The thing about track days is that you have the luxury of taking your time if you want, which also means you have the luxury of saving your car on the straight, too. So why not use it? There's no rule that says you have to squeeze every last rev out of your engine, so you might even think about keeping a few revs in hand. It may well rev to 7000rpm max, but if you're going to use it for your daily transport in the week, then maybe give yourself a rev limit of 6000. It will save the engine, and all the fun is in the corners anyway. Maximum revs is not really the way to go, anyway, as you'll find that maximum power will be a little before the peak revs, and the car will perform a lot better if you rev to this point.

One frustrating thing about road cars on a race track is that you can find you haven't a gear to suit a particular corner. The idea is to be in a gear that allows you to drive the car through the turn. Now, if it's too low a gear the revs could be too high, or you might not take enough speed through a corner, while if it's too high a gear the engine could get bogged down with too few revs. In many racing cars you can change the gear ratios, and the object is to get to the point where you shift into the next gear just after the exit of a corner, but it's a bit more of a compromise in a road car.

If you do find a corner is in between gears – and you will – then you should perhaps choose the higher gear rather than scream it through with the lower cog, which will result in harsher weight transfer when it comes to lifting off, or maybe wheel-spin in the corner.

You should also try to avoid changing up in the middle of a corner if you can. If it is unavoidable, and it might be at some circuits, then try to make sure you've plenty in hand for it. Remember that as you lift for the change there will be a sudden weight transfer that will take some grip from the rear.

Actually, you'll soon find that there might be a corner on every track day which isn't right for your car, but just learn to live with it and enjoy the rest of the lap. Accepting the compromise is a big part of doing track days in a road-going car.

Changing down

One of the first bits of advice you might be given at a track day is to change down towards the end of your braking for a corner. The reason for this is to not upset a car that is already braking to the maximum at high speed with the extra weight transfer through the deceleration of the downshift.

As to whether you go down through the gearbox, say 4-3-2, or simply skip to the gear you require, then it seems to depend on what you prefer. That said, most instructors will tell you to go down through the gears, simply because it's more progressive and you've less chance of finding a gear without having completed the required amount of braking.

Either way, you need to be very careful with your downchanges. If you get them wrong – hook second instead of fourth from fifth, for instance – you can have a very expensive over-rev on your hands, and the rev limiter won't save you this time, because you've fooled it by going straight to the lower gear, finding those big revs through the back door. Oops.

Oh, and if you're serious about track driving, it's well worth learning to heel and toe.

Heel and toe

I've always thought that *The Heel and Toe* might be a good name for a motor racing themed pub. But one of the arguments at the bar might be about what exactly it meant, because it's a term that can cause confusion. Think of it more as 'side of foot and toe' and it will be a little easier to understand – though that's a rubbish name for a pub.

The idea is to slide the side of your right foot on to the throttle, to 'blip' it, while still keeping the ball of your foot on the brake as you're slowing down. All this happens while you're changing down at the same time, as the stick passes through the neutral position, with the blip coming just before you take your foot off the clutch.

But why bother? Well, when you're braking to the maximum the last thing you want is to apply extra braking pressure to slowing wheels, as you're at the maximum anyway, so any more braking will cause the wheels to lock. But if you're changing down at the same time this will bring on a degree of engine braking, and it's this extra braking that can cause lock-ups – stand by a hairpin at a track day and you see it, and hear it, all the time. There's also the extra weight transfer from catching the lower gear which can unsettle the car, not to mention the 'driveline snatch' from

the sudden spike in revs. None of which is good.

As we've said, the best way to get around all this is to flick the side of your foot over to the throttle and blip it as you make your down changes – the heel and toe. This blip, if timed right – and that's important – will bring the revs up to match the lower ratio. Practise it first, and if you find that the car seems to lurch forward, then you've blipped too much. While if, on the other hand, the revs spike and the nose dips, then your blip wasn't big enough. One of the great things about this technique, incidentally, is that it is one of the few on-track skills that can be honed on the road without annoying the local constabulary.

In some cars the distance between the brake and the accelerator is far too great to make this an easy process. If that's the case with yours, then you can usually buy special pedal extensions from most go-faster goody shops.

But why is it called heel and toe then? Well, that comes from the good old days before the Second World War, when the throttle in a racing car was often a dodgem-car-like hinged set-up with the brake pedal much, much higher, to allow for the ropey anchors those heroes used to race with, so they had to twist their heels to hit the accelerator. Often the pedals were reversed, too, so that the throttle was in the middle, which gave

BELOW Blipping the throttle as the shift passes through neutral on the downchange – heel and toe – will help balance the car in the braking area.

David Coulthard a lot to think about when he tried a Mercedes W125 out for size a few years ago; or so I've heard.

In the past, heeling and toeing was always taught with double-declutching, where the blip came as the stick passed through neutral but as the clutch pedal was out, so you had to pump the clutch twice flicking your foot over to blip while the pedal was up and the stick was in neutral. Actually, that's not as difficult as it sounds once you've got used to it, and it's the process I was taught at the Winfield Racing Drivers School in France back in the early 1980s. These days there seems to be little point to it, to be honest, though some American text books and race schools still teach it.

It's definitely slower than single clutching, though. That said, on older cars it can be easier on the gearbox, and if you do it right it's a smooth way to skip gears – fifth to third, for instance – but any advantage you gain from it is undone with the time it takes. Whatever, it's very hard to get right, and once you've got it so it's second nature then it's even harder to stop doing it! I for one just can't shake the habit.

Incidentally, being used to flicking your foot across to the throttle when it's on the brake has another advantage, for if you do get into trouble and it's both feet out on clutch and brake, you can keep the engine running with a couple of well-timed blips.

Dead time

One thing you should try to bear in mind out on track is that there should be no *dead* time when it comes to the pedals. By dead time I mean your right foot needs to be either on the accelerator or the brake, except for the transition time between the two, of course. The whole point is to go quick, so you should always be slowing down, cornering hard or – most of all you hope – speeding up.

So, assuming we're straight line braking for now, as you've finished coming off your brake and you've turned the car in, which should be pretty much at the same time, you need to be smoothly rolling you foot from the brake and on to the throttle. You don't want to be accelerating just yet though; the tyres will have enough to cope with, with the hard cornering that is the first part of the turn.

You'll still need a little throttle, but this is a part throttle opening that is neither accelerating nor decelerating, so that you still have some drive going through the driven wheels and consequently control over the car. It also gives you a chance to trim the car's balance if you're understeering or oversteering.

This is called a 'balanced throttle', although sometimes people will call it a 'trailing throttle', which is slightly confusing as this can also mean coming off the throttle to transfer weight off the rear – a technique you can use to dial out understeer.

The trade-off

There are three phases to a corner, which you might label the braking, cornering and accelerating phases. But, as we've already seen in our look at the line, there's a fair amount of overlap involved in the latter part, where you'll need to both steer and accelerate at the same time.

Think of it this way. If a tyre is using 100 per cent of its capability in clinging to the road

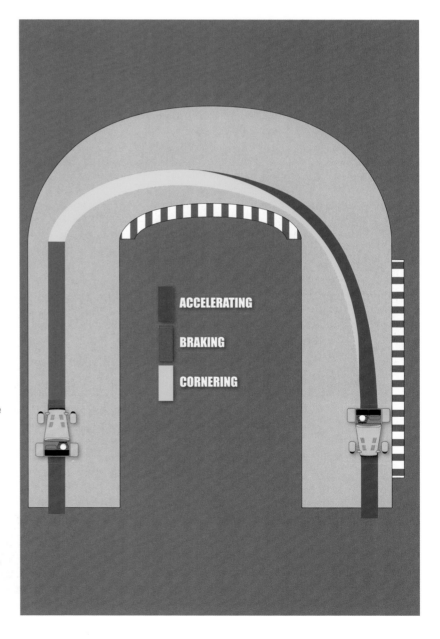

BELOW Fig. 13. Note how as the driver feeds in the acceleration he is also straightening the car, trading off cornering for straight line speed to exit the corner as fast as possible.

ACCELERATING

BRAKING

CORNERING

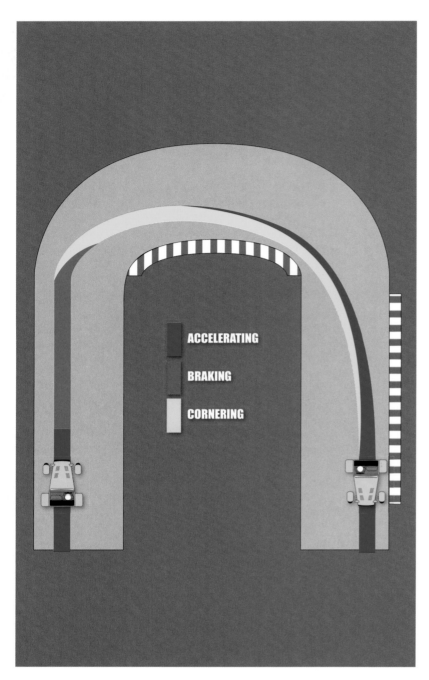

ABOVE Fig. 14. Trail braking is about progressively coming off the brake as you go through the first part of the corner. It can help, as it keeps the weight over the front while you're turning in, but newcomers are better off straight line braking to begin with.

to the wheel, and I reckon that's a good way of looking at it – more gas, less lock.

But this can apply to the brake pedal, too, especially in the slower corners. You see, as we alluded to early on in this chapter, it's not just out of the corner where you can overlap a pedal movement with cornering – you can also keep on the brakes into the turn. This is called trail braking. But be warned, this is certainly not suitable for all cars, and you really have to have a good understanding of what the rear of your car is doing before you start to experiment with it.

Basically this involves *bleeding off* your braking as you go into the turn, so that there's a real overlap in the braking and the first part of the corner. One of the reasons for this is to keep the weight over the nose during the turn-in so that the front of the car does not go light when it needs grip.

But you still need to think about the trade-off here, because you cannot expect to brake at 100 per cent and turn at the same time. And that's why the expression 'bleeding off' is italicised above. So, you ease off your braking a little and feed in the steering, trading off the two into the turn in the same way as you trade off between throttle and lock on the way out.

During 2007 I wrote a series of features for a magazine in which I talked to a selection of top drivers from various racing disciplines from the past and present, and it was clear that most of them trail braked to one extreme or another in the slow corners. When it came to one well-known British Touring Car driver, he told me he was braking – or rather easing off the brake – right up to the apex. But then he had a car that was designed for this. Your road car might not be, and in fact it probably isn't. If you do think it will work for you, though, experiment where there's run off – ideally at an airfield venue – or get an instructor who is familiar with the technique to talk you through it.

in cornering, then there's no way you can add even another one per cent to the equation – accelerating or braking – without that tyre breaking away into a slide. Which is why, as you add the gas, you take away the steering angle – unwinding the lock from the apex – balancing the acceleration against the cornering.

This is another reason why it's important you are progressive on the throttle out of a corner, and you really should try to think of it in terms of easing on the accelerator *as* you're unwinding the lock. A veteran racer once told me that he always tries to imagine the accelerator pedal is somehow linked

Driving a corner

Right, time to put all this stuff together and drive through a corner; we'll use a third gear right-hander that leads on to a long straight as an example and, to keep it simple, we'll do all our braking in a straight line.

You're into the braking area, focused on the turn-in bollard, and judging your braking with the aim of finishing it as you reach this point. First brake input is soft, and then you brake progressively harder until you're braking at the maximum, the point just before the brakes will lock (which is called threshold braking, incidentally).

The nose of the car is dipping now the weight

is on it, and as you come to the end of your braking you make two heel and toe downshifts from fifth into third, while you're also sighting the apex bollard and looking beyond it to the exit bollard, visualising the line you'll take from the turn-in to the exit, via the apex.

As you reach the turn-in you're progressively coming off the brake, coming right off the pedal as you turn. The turn in is smooth and you're winding on just enough lock to get you to the apex, and now the weight starts to travel across the car, while you're also rolling your foot over to get on to a balanced throttle.

So you're into the corner, and the weight is transferring across the car, until there comes a point where all the transfer that's going to take place has taken place, the car has taken a 'set', and it's balanced enough for you to begin to drive it through the corner.

You clip the apex, and now you're progressively feeding in more and more throttle while, at the same time, winding off the steering lock to straighten the car, until you're at maximum acceleration and the car is straight, hitting the exit of the corner right at the very edge of the track. Nice one.

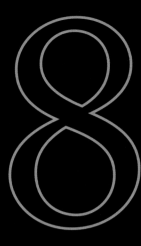

Car control

Car control

It can be great fun to slide a car at a track day, but sometimes a slide will take you completely by surprise – the secret it to think through what actually went wrong and to learn from it.

Hanging it out

Right, by this stage I guess you're getting a bit bored with all this talk of keeping it smooth and safe, and what about sliding it about a bit? Fair point; after all, track days are supposed to be about having fun, aren't they?

That said, everything we have talked about in the previous couple of chapters will give you a safe base from which to start to experiment with your driving – and then the fun really begins. But you do need to get the basics right first. That way you'll begin to understand the way the weight works across your car and the balancing act you're doing with the tyres to trade off – at the very least – cornering and acceleration.

Understand what's happening to the car and you'll be able to induce little slides by backing off in the corner, then putting the power on to pull it back into line, or maybe if you're in a rear-wheel-drive car, putting a bit more power on to come out of the corner in a power slide. Huge fun.

But a word of warning here. Many track day operators take a dim view of someone getting rally car sideways on every corner, and you could even be black flagged. But that depends largely on whether the marshals reckon you look like you're in control, or if you look like you're rushing towards the scene of your very own accident.

You'll probably find that small slides will give you thrills enough anyway. Hang it out too much and you'll just scrub off speed and ruin your tyres – and that never feels good.

Moments

Drive on track for long enough and it's almost inevitable you'll overstep the mark at some point. Now this doesn't necessarily mean you'll crash, beach yourself in the gravel trap, or even spin. More likely it'll simply mean you'll have a 'moment'.

A moment is a pocket adventure: the car snapping sideways, a wheel locking up, sliding on to the grass. In a race it's something you need to put behind you, just get on with the job in hand, but at a track day you have the time to learn from it, and that is precious.

So, if something happens that you were simply not expecting the best thing you can do is pit and have a good think about exactly what you did to cause the situation. Even better is to talk it through with one of the instructors if there's someone available. But just make sure you think it through, or talk it through, while it is still fresh in your mind.

Go through in your mind exactly what happened, remembering how the weight transfer will affect the way the car handles. Chances are you'll soon have a clear notion of what happened, and you can get back out with confidence.

'Oppy locky'

The crucial thing to remember when your car gets out of shape is that you really *must* look towards where you want to go rather than towards where the nose of the car is pointing. If you get it very sideways and you have plenty of opposite lock on, this might mean you're almost looking through the side window, but that's far better than looking out the windscreen in this case. People have a natural propensity to follow their noses, so if you do find you're looking along the nose of the car towards the inside of the turn, then there's a fair chance that's where you'll end up. Looking where you want to go will also help you when it comes to winding on the right amount of opposite lock,

FAR LEFT When the car gets out of shape you must remember to keep looking in the direction you want to go. Oh, and yes, he did save this one.

SEQUENCE – CONTINUED OVERLEAF It's very easy to collect a slide and then have the car snap the other way, especially in the wet, but keep your eyes on where you want to go and there's every chance you won't over-correct.

without over-correcting, which can simply cause another slide in the other direction.

There are plenty of other ways you can get it wrong on a race track, but one very common mistake newcomers will make is to turn in too early, which leads to a premature apex and then a lack of road on the exit. Usually by this stage he's also got loads of understeer as he winds on more lock to make the turn. In the worst case this ends with a sudden lift off the throttle and then a spin from the equally sudden transfer of weight.

Another rookie problem is simply going into a corner carrying too much speed. If you find yourself in this situation the best thing to do is to stay on those brakes rather than trying to make the turn – it's a track day, not a race, remember. But if you've started to turn, then wind off the steering lock so you can brake more efficiently. Now you might effectively be steering towards the outside of the turn, but that's OK. Just wait until you feel it is slow enough to turn in, then try again – but take into account all the slippery rubbish that can gather off line.

Offs (gulp...)

If you've spun, then the chances are – on a well-run track day at least – you'll be called in to have a chat with someone. It's not so much a telling off (though it could be if you make a habit of it and end up taking time from everyone's day) but more of a debrief, just part of the never-ending process of learning about driving fast on track.

Yet while reporting to teacher is all very well after the spin, what should you do *during* the spin? After all, most crashes start their brief but spectacular lives as humble spins. The best advice, especially for the beginner, is both feet out – declutch and brake – and also make sure you've a light hold on the wheel, while if you've hooked your thumbs over the spokes then make sure they're now on the rim of the wheel.

By braking and declutching you're at least likely to scrub off some speed, while if the spin has occurred on the way into the turn you may even increase its radius and stay close to the racing line where the grip is. More important, though, by getting the car all locked up you should keep it going in some sort of predictable direction – predictable to you and to those on the track with you.

One vital point, though, and that is you should never let go of the brakes while you're still spinning, however slowly you think you're going, as you can easily misjudge your speed and pitch yourself into another moment, a spin in the other direction, or even the gravel.

LEFT Gravel traps do an excellent job of stopping errant cars before they hit the barriers, but try to go into them with your wheels straight, that way there's less chance of the car digging in and maybe even rolling.

BELOW One downside of gravel traps is that you'll have to spend ages cleaning all the gravel from under your car, and especially from around the brake calipers.

These days many spins do tend to end in the gravel trap – highly effective run-off areas filled with gravel that bogs the car down. Once you're in the gravel it's pretty much game over, so switch off and wait to be rescued. But don't get out of the car – this also applies if you've come to a standstill and can't restart on the circuit. Wait for the red flags to fly and rescue vehicles to arrive before you abandon ship, and only get out when you're told to do so – and this goes for your passenger, too. Of course, if the car's on fire then ignore all of the above and get out as soon as you can. But be careful to keep your eyes on the direction other cars are coming from, even if the red flag's flying, because you can't ever be sure everyone will have seen it; and make sure you get well away from the track, and preferably behind a barrier.

If you're unlucky enough to have a car spin in front of you, then the best advice – for the novice at least – is to hit the brakes. It might cause you to spin in sympathy, but at least it will slow the car. And if your car does spin as a result, well it's still better than trying to second-guess the path a spinning car will take – go to a race meeting and see how often even the professionals get that one wrong. Incidentally, there's one piece of racing folklore that says that to avoid a spinning car you should aim for wherever it has already been, but that's a hit and miss approach – quite literally.

As you become more experienced in track driving you'll begin to get a feel for how a car

ABOVE If a car spins in front of you, the safest approach, for a novice at least, is to get on the brakes – trying to predict its path can just cause more problems.

will spin, and you'll also start to understand how you can miss it, maybe by lifting off, catching the resulting slide and throttling through the gap; but to begin with, for most emergency situations think 'both feet out' and get on the brakes.

Except... nothing's quite that simple, and there'll be times when getting on the brakes will be the last thing you'll want to do, such as when you've just run wide on to the grass. This can happen quite easily when you begin to push, but at most circuits it shouldn't really create too much of a problem as long as you're gentle with the car. The secret is to *ease* the car back on the circuit. Keep the steering as straight as possible and keep a light hold on the wheel, then gently nudge it back on track.

Whatever you do, though, don't lift off the throttle suddenly. This is vital. You need to just ease off the accelerator pedal a little, as backing off completely will just cause a massive weight transfer which will unbalance the car on what can be a slippery and uneven surface.

In these sort of off-track excursions it's often all about keeping the wheels as straight as possible, and that is also good advice if you go straight on at a corner to the extent that you're off the track.

Keep it straight and there will be less chance of the wheels 'digging in', which could cause a roll. Remember, you're not in a race so there's no rush to get back on track; you're much better off bringing the car slowly to a halt.

The same advice goes for those gravel traps; try to go in straight if you can to avoid the risk of the turned wheels digging in. And be prepared for the tedious job of picking out all the bits of gravel later on in the pits – it'll get everywhere, including in the brake calipers, so you need to be very thorough when cleaning out the gravel after an off.

Obviously, then, there's a lot that can go wrong on a track day, and we've only scratched the surface here, but as long as you remember to watch your speed into corners, to unwind your lock as you accelerate out of corners, to treat the accelerator like a control with a wide range of settings and not a switch, and to give yourself time out to think through moments, you should be OK.

In fact, actual crashes are *extremely* rare on track days, although at some circuits there's obviously more chance of getting closely acquainted with the scenery than others. Thing is, as you become more experienced, thinking

through why something happened after every moment or incident, then you'll start to *anticipate* problems rather than reacting to them, and this is obviously better. So to start with, as indicated at the beginning of this book, you might do well to make your mistakes at airfields rather than at circuits – just don't forget to learn from them.

Fighting the systems

One recent complication to the whole track driving process, especially if you plan on using your road car, is the electronic systems, stability controls, and all sorts of other gizmos. Now, these are quite wonderful systems on the road and they have no doubt contributed hugely to road safety, but on the track they can cause problems. Not so very long ago I was testing a brand new front-wheel-drive car from a major manufacturer on a circuit, and after a couple of laps getting up to a speed I found that I was impressed with its capabilities, particularly its stability and grip, but it was soon time to play, just to see what happens when the back steps out. So, coming into a favourite corner I turned in on the brakes, knowing there was plenty of room for me to throttle out of the slide.

But there was no slide, or at least not the slide

I was expecting, as the car simply straightened itself and speared over the harsh inside kerbing, which then launched me across the track and over the even harsher outside kerbing. It simply didn't compute – or rather it had *computed*, as I soon realised I had forgotten to switch off the stability control button.

Not all of these systems are bad, indeed some are pretty good, and to be honest if I hadn't been fooling around I wouldn't have triggered the system anyway, but if you do find the 'safety net' on your car is a little intrusive, then you'll need to make a point of switching these systems off – if they can be switched off – every time you start the car, as they usually come on automatically as a failsafe. So make switching it off part of your pre-flight checks. Increasingly, though, even with these systems switched off the manufacturer will make sure there's still an element of stability control to save us from ourselves, it's just the way the world is going, sadly.

If you are forced to track a car with such a system then you should really remember that some of the normal rules might just not apply. The system is there for the sole reason of making sure you do not spin, to the extent that in the worst

ABOVE When a car slides wide and on to the grass you need to try to keep it as straight as possible and then ease it slowly back on track.

FAR RIGHT Left foot braking can be of use in front- wheel-drive cars – like these Clios – on a wet track, and even in the dry, but it takes a lot of practice to get it right.

case scenario it will just slow you down while making sure you go into the gravel – or barrier or other car – head on. The problem is if you do all the normal 'driver' stuff, chances are you'll be fighting against a system that's far cleverer than you. Who knows what sort of mess you could get yourself into?

The best approach is to study the manual, find out exactly what nannying devices you have on your car, how they work, and when they will kick in. After that, go to an airfield for your first track day in it and start to experiment. Better still, buy an old car instead!

Not all driver aids are there to spoil track days, though, and ABS (anti-lock) braking can be a great advantage, and while it might not actually be more efficient in the braking zone than a good driver – a long-running argument – its advantages in the event of an emergency cannot be sniffed at. But the important thing with ABS in an emergency situation is to *use* it, do not be put off by that awful juddering sound or the 'kickback' through the pedal as the system comes in.

As for traction control, you're never going to learn to be a truly good track driver if you're leaning on it all the time, but when it's wet you might be very happy you've got it, as we will discover in the next chapter.

Left foot braking

It's worth briefly going into left foot braking here; simply because you'll hear more and more people talking about it at track days. Left foot braking is a technique that's vital if you're at the wheel of a front-wheel-drive rally car on a loose-surface special stage, but can also have its uses on the race track. In a rally car you brake with your left foot to cut down the understeer – it allows you to keep some throttle on to stop the fronts locking, which also means the rear wheels are being braked more, bringing the back around – and there's possibly an argument for using a similar technique at tight turns on a wet track. That said, it takes lots of practice to get this right – there's even a well-known school entirely dedicated to teaching this technique (see Appendix 2).

Where left foot braking is probably more of a benefit for track drivers is in fast corners, where a quick dab of the brakes can settle a car into a turn while the right foot's keeping the revs up. Though, to be honest, a lift of the throttle is probably better in most cases anyway as it will still shift enough weight on to the nose and you can then accelerate the car through the turn. That, though, depends largely on the car and the corner.

Where many argue it's of particular benefit to *race* drivers, though, is in the way it cuts down the time it takes to get your foot from throttle to brake, both in terms of reacting to a situation and in speeding up the braking process.

Here are a couple of examples of how both might work. First, reaction time, and this is only of benefit if you have your left foot hovering over the middle pedal – something you can do if you ultimately right foot brake at the corner. In this case it's just a split second movement to get on to the brakes should a car in front suddenly slow, rather than a lift of the right foot off the throttle and then moving it on to the brake. This makes as much sense on the M25 as it does in a pack of closely matched cars in a race, but on a track day you shouldn't really be following other cars closely enough to get caught up in their problems, so it's probably a little less relevant.

As for speeding up the braking process, here it gets a little more complicated, chiefly because you still have the gearchange to worry about. Left foot braking track drivers – a rare but growing breed – will argue that the time spent moving your right foot from throttle to brake as you come into a corner is 'dead' time, though again this is more relevant to shaving tenths in a race situation than for track day driving.

That said, it all makes perfect sense in some of the new breed of road and track cars with sequential changes which don't require you to use the clutch, but it's a little more difficult if you want to change down in a regular way or, indeed, heel and toe.

Clutchless gearchanges are possible, if you match the revs correctly, but you're always going to get that wrong a few times an hour in the heat of lapping, and nasty gnashing noises could be just the start of your problems. With that in mind, perhaps the best advice is to fit an over-large brake pedal – maybe one from an automatic car – which has the room for both the right and left foot at the same time. With this in place – and using a single down-change turn as an example – you would brake with your left foot, then move the right over to the big brake pedal while the left is still braking, then move left to clutch, downchange and blip with right – while braking, too – then move left back to the big pedal, then right back to the throttle. Phew!

There are many advantages to left foot braking, but it's a *very* difficult art to master, particularly if you've spent much of your life right foot braking on the road, so be very careful if you feel the need to experiment.

9

Wet weather driving

Wet weather driving

There's little that can match the challenge of a wet track. But that doesn't mean driving in the rain has to be dicey, as there are wet weather techniques that should help you lap both safely and rapidly – while having a whole lot of fun in the process.

Some love driving in the wet, some hate it. Some will even pack up and go home when it rains, or spend the time when it's really pouring drinking tea in the paddock café, glancing at their watch every now and then, as their track time dribbles away.

But you need not let the rain spoil your day. OK, there's an argument that perhaps you should simply not venture out on track if you're not comfortable with the conditions, but there's another point of view that's worth listening to as well, and this is that if it does rain during your track day then don't treat it as a disaster, treat it as an opportunity. An opportunity to brush up on your driving skills and car control in an environment that demands you get the basics right. So, learn to feel comfortable in the wet, learn to go well in the wet, and learn to love the wet.

Smoothly does it

A wet track amplifies everything you do in the driving seat. Throttle, brake and steering inputs that are too heavy or too sudden can result in an instant loss of traction, a lock up, or even a spin. It's not just the obvious things, either; something like a snatched gear change can also cause problems, and even just letting out the clutch viciously can result in a snap in the driveline that can easily lock the wheels when the grip is low.

Remember, everything you do in the car is transferred through the tyres to the road, so when it's slippery and those tyres are not working so well, you need to calm it all down. You should be aiming for smoothness, anyway, but now you'll want to be even smoother, and almost gentle in the way you use the controls. That is why it's even more important that you're relaxed at the wheel than it is in the dry, with none of that white-knuckle gripping of the tiller. Problem is, a wet track tends to make people nervous, and a nervous driver will strangle the wheel. Tough one that, but one you can work on with experience.

Braking distances will, naturally, increase in the wet, while your cornering speed is going to decrease, to the point where you may well be taking corners at least one gear down on the cog you used for dry weather lapping. That said, a higher gear at the same speed but using less revs will mean less chance of wheelspin because there's less torque acting on the wheels, so there's an argument for that, too. But the great thing is you're not in a race, so you can experiment, build up your speed slowly, and try a variety of approaches.

Another thing to be careful of is judging the amount of grip that's available to you by the way the car in front has taken the corner. It may be a slower car than yours on paper, but on a wet track the spec books get a bit soggy, and often less powerful, more agile, cars will be better.

Then there are the tyres to think about; sometimes ordinary road tyres will cope much better than full-on track tyres, while a car with soft road suspension will almost certainly be easier to drive than a stiff race-style set-up. Also, heaven forbid, the guy or gal in front may just be a better – or at least a more experienced – driver than you. Point is, don't trust to others when it comes to your approach to corners in the wet.

Driving in the wet is a much more personal challenge anyway. It's a very *feely* sort of driving experience. You need to *feel* out for the best possible grip that's available, and that's very often not to be found on the regular racing line.

The wet line

Experienced track drivers will often take a different line in the wet from the one they use in the dry. This is for a variety of reasons: perhaps a clipping point is now in a deep puddle of water, or maybe there's a bump in the track that's no problem in the dry but best avoided when there's a chance you might lock a wheel or suddenly lose traction, or there might even be a little rivulet running across the track at your turn-in point. But the main reason is that on a working race track a lot of rubber and other deposits are laid down on the track surface over time, and this will naturally be where the cars pass over it – on the line – and when it rains the water will sit on top of all that rubber, making for a very slick surface.

This is why you might see cars taking unusual

FAR LEFT A wet track demands that you are smooth with your driving and almost gentle in the way you use the controls.

BELOW The best grip will not always be on the regular racing line, and you'll need to avoid the wetter parts of the track.

ABOVE When the rain stops, a dry line will begin to appear where the cars pass over the same part of the track.

lines in the wet, sometimes a car's width and more off the regular line, and maybe even around the outside of a corner. Of course, if you do this you'll still need to cross the regular line at some point, both in and out of the turn, but you now need to be aware of it and modulate the throttle accordingly and, where this happens, you might need to try to keep the car as straight as possible.

The wet line works, too. To see that, you only need to watch footage of the real racing greats of the modern era (drivers like Michael Schumacher) when they're racing in the rain. Often they will be far from the regular racing line as they sniff out the available grip – and far in front of the other drivers, as well.

However, just because you've found a line you're comfortable with, that's not the end of it. A track is a living thing, it's always changing, and in the wet it lives fast. As a track gets wetter, or as it dries, the line can change. So remember, you should always be feeling for that grip.

Feeling for it, and looking for it. Sometimes especially glossy sections of track can be treacherous, while a stretch with a grainier, darker grey look to it will often offer up more grip. Not all parts of the same circuit will react the same to the rain, either. Indeed, at some circuits one section

of the layout may offer far more grip than another – at Silverstone the Brooklands-Luffield-Woodcote section, for example, can be like ice when it's wet, while on the same day Copse will give you a surprising amount of grip.

The track surface may well be more treacherous after its first soaking, when oil and rubber deposits float to the surface, but a drying track can also cause problems, as a dry line will start to appear where the cars have been drying the track by constantly running over the same bits of asphalt. This line can be particularly tricky – a high speed high-wire act, in fact – and it calls for great precision from the driver; for you only need to drop a wheel on to the wet stuff and you could get pitched into a spin. Actually, at a track day, maybe this is the best time to take a breather, and let the others do the 'drying up'.

Something else you need to be careful of in the wet is painted kerbs, and all the other painted parts of the track surface. These can get very slippery indeed, and while a wet kerb might, arguably, be a risk worth running in a race, there's really no point on a track day. Similarly, white lines are notorious in the wet – just ask one Nigel Mansell about Monaco '84 – so be careful to avoid them, especially those that mark the track

edge in the braking area of a turn; put an outside wheel on one of them and it's instant lock-up. If you do feel the need to ride the kerbs in the wet, make sure it's on the slower corners and it's with your unloaded – inside – wheels. But the best advice is simply to avoid them.

Self help

Even if you're avoiding the white lines and the kerbs, there's still a chance of locking up a wheel under braking. Unless, that is, your car is fitted with anti-lock braking (or ABS). Now, opinions are divided on this bit of kit, which is fitted to many modern road cars. Some see it as a lifesaver – quite literally – while others see it as an intrusion, something that dilutes the purity of the driving experience.

Whatever your views on ABS and other aids, such as stability control and traction control, you'll have to admit that when it's raining they're a great help – just not quite so much fun. The problem with fun, though, is that there's always a downside, and the downside to having no driver aids – as will be the case with many older road cars and most bespoke track cars – is you're on your own. Which basically means it's up to you to be your own ABS, and your own traction control.

As far as being your own ABS is concerned, this will mean learning the art of cadence braking. This involves modulating the brake pedal when the wheels lock under braking. Basically you're pulsing the pedal in order to retain some sort of steering control. So, the front wheels lock under braking, then you're off the pedal, applying the steering wheel when there's grip, then back on the pedal for retardation again. It's modulating between the steering movements and the brake pedal inputs, making sure that you can both slow down and steer the car in turns.

ABOVE This driver has allowed his Caterham to drift wide on to the kerb – but these can be very slippery in the wet so it's a risky practice.

BELOW A drying track can be a tightrope act, and it's important not to drop a wheel on to the wet bits.

ABOVE The white lines at the edge of the track can also be very slippery in the rain, and they are best avoided.

Avoidance braking is similar. The problem you're going to have when your brakes lock up is simply going straight on, directly to the scene of the accident – be it the gravel trap, the barrier, or a car that's spun in front of you. To attempt to avoid the hazard, though, you can turn the steering on to quarter lock, which to begin with will have no effect whatsoever because you're all locked up. Yet, when the moment's right – before you've hit the hazard and when the car has hopefully slowed a little – you can come back off the pedal and you should suddenly have enough steering capability, because the front wheels aren't now locked, to avoid the problem. This takes practice though, and a visit to a skid pan is a great way to master this and other wet weather driving techniques.

Throttle control

You also need to be easy on the accelerator when it's wet. Assuming you've no traction control – and even if you have you should still be careful – then throttle inputs should be gentle rather than harsh, almost speculative in fact. Remember that you're always feeling for the grip, and your fingertips in this respect will be the steering and the driven tyres. In fact, you should almost treat the throttle pedal like it's made of fine porcelain. By the way, accelerators with a lot of range in their movement from off to full throttle are always better for wet weather driving because they can be modulated more, meaning you'll have far more control – the last thing you want in the wet is an on-off switch for a throttle.

As mentioned above, some track day drivers, particularly those in turbocharged cars, will take

FAR RIGHT Standing water: when it gets this bad you just have to try to avoid it – either that or risk an aquaplane.

a corner a gear up on the norm to prevent the power coming in in one big lump. Indeed, Ayrton Senna used this technique to good effect while winning his first grand prix in monsoon conditions at Portugal in 1985.

Similarly, short-shifting can make sense in the wet, grabbing a gear before the engine reaches the revs where it's at maximum torque can mean the power delivery is a little softer and there's not such a surge of power. Basically, you're trying to avoid wheelspin as much as possible.

But what if you've trodden on the loud pedal just a bit too much – and here we're assuming you're in a rear drive car – is there anything you can do to get the car back on the straight and narrow? Well, the obvious thing is to lift off the throttle, and that's pretty much a natural reaction – we humans have an inbuilt cause and effect chip – but you should be careful not to lift off completely, and if possible you should also try not to lift off too suddenly. You need to try to be as progressive as you can; you really need to just *feather* the throttle.

The reason why you should not release the accelerator completely in this sort of situation is that you want to keep a degree of forward motion in the car. Quite apart from the sudden weight transfer to the front of the car, which will take the weight off the back, you'll also have a degree of lateral movement from the slide. So, if you're sliding viciously to the left out of a right hand turn, the car is likely to spin to the left. But, if you've kept on a little accelerator – let's say you've slackened it from 90 per cent power to 50 per cent power – then you've also kept the forward motion to pull you out of the spin.

Aquaplaning

When it comes to aquaplaning, the advice is just a little different. Now, chances are you've experienced an aquaplane on the road; it's that horrible moment when you hit a patch of standing water and the tyres lose contact with the asphalt and you seem to be skimming across the surface like a hovercraft. It's a result of the tyre treads being unable to disperse the water, and because of this it builds up beneath the tyres, lifting the treads away from the road surface.

The first thing you need to do in this situation is make sure the steering is straight. You really do not want the front of the car to come out the other side of the standing water and then hit grip with some steering wheel lock wound on while the rears are still aquaplaning, that will spin the car in a heartbeat.

Ideally you don't want any lateral force going through the car when you enter the aquaplane either, as an aquaplane will obviously accelerate that movement, and around you'll go. So, even in a corner, you should look at trying to straight-line standing pools of water if possible – assuming there's no way of avoiding them altogether – picking up the steering again once you're out the other side.

You should also try to go through the aquaplane on a constant throttle – although you'll obviously feather the throttle a little because the revs would have shot up when you hit the water and lost grip. But you *must not* touch any of the other pedals. Then, when you come out the other side you should be able to pick it all up again without any drama; or at least that's the theory.

The thinking behind this is that if you've lifted off in the aquaplane, then when you come out the other side the engine speed – and hence the speed of the driven wheels – will not match the actual speed of the car. In this situation there's a high chance of a spin.

All that said, aquaplaning is very difficult to deal with. Even Formula 1 drivers get caught out by it. So, if there's standing water out on the track, perhaps this would be a good time to sit out a session and have that cup of tea.

Building up speed

One other tip when it comes to wet weather driving is persuading yourself to start the day all over again when it begins to rain, and this is particularly pertinent when it comes to session track days. On these days most will want to maximise their track time, so if the rain starts when they're out on track there's a fair chance they won't come straight back into the pits. If you find

yourself in this position then by all means stay out, but adjust your speed *significantly*. The best advice here is to slacken off your pace back to just one-tenth – assuming you were somewhere close to ten-tenths, or the limit – and start building up your speed gradually again, remembering that a track can be at its very slippiest when it gets its first soaking. In other words, treat the onset of rain like it's a brand new track day. It is.

Incidentally, some race drivers talk about psychological rain, which are the drops of rain you see on your visor or windscreen before it has properly started wetting the track. For them it's important to ignore it, and drive through it until the track is properly wet, but then they're in a race. You're not in a race, remember, so err on the side of caution, and when it starts to rain, ease off.

Visibility

While being seen is important – most operators will expect you to run with lights on in really poor weather – it's what you can't actually see that causes most problems in the wet at track days. This, though, is nothing compared to what's encountered in races, particularly with single-seaters where huge rooster tails of spray will hang off the back of the open-wheeled cars. It calls for the development of special skills, and the use of all the senses, to go well in the pack during a very wet race.

Thankfully, at a track day you need not run in a pack of cars, and you certainly will not want to when it's wet. It's always important to give other cars space out on track, but when it's raining this is doubly so, for not only is there a greater chance of slipping up, there's also less of a chance of stopping in time to avoid someone else's slip-up.

If you're running in a regular road car, some of the problems many race drivers face when it comes to visibility simply should not be much of an issue. But if you're in a special, a modded car, or an ex-race car, then things can get tricky.

This is chiefly because of misting up, whether that's the windscreen or the visor. In the first case, many drivers will rip out demisters in order to lighten their cars, only to find the windscreen steams up once it rains. So it might be worth bearing that in mind when you're thinking about what modifications to make – either that or take a squeegee or a cloth in the car with you; many do.

One way you can stop your windscreen from misting up in the first place, though, is by trying to stay as dry as possible when you're out of the car. It doesn't take a genius to figure out that if you stand around in the pits getting soaked, once you climb into a warmer car the water's going to turn to steam and mist your screen.

If you're in an open car, though, there's the problem of your visor misting up. Now, this problem has exercised the minds of racers for years, and everyone has their own particular theory about how to cure it. One old favourite is to smear a film of washing up liquid on the inside of your visor, but perhaps even better is a small block of tape that stops the visor closing completely, while many helmets will have ratchet systems so the visor will stay up just a little bit anyway. Then again, there are plenty of chemical solutions on the market for both the inside and the outside of the visor, often available from racewear or motorcycle shops.

Other fashion accessories for the wet include a good coat for when you're not in the car, particularly if you're doing a lot of airfield days, and maybe a tarpaulin to cover all your gear when you're out on track – this is especially important if you haven't access to pit garages.

Karting rain suits are a great help if you drive one of those naked track specials, such as the Aerial Atom, while I've even seen people driving in electrically heated rain suits designed for bikers – just the thing for an open-topped car on a winter track day.

As for footwear, it goes without saying that you should try to keep the soles of your shoes or race boots as dry as possible – a foot slipping off a pedal could prove disastrous. There are even special overshoes, that slip over your regular shoes, available through the racewear specialists. That said, you don't really have to go to all that trouble, and even Ayrton Senna once advised that you should wrap plastic bags around your boots, and if it was good enough for him...

BELOW Most operators will ask you to lap with your lights on when conditions get very bad.

Rain check

As far as car preparation for the rain is concerned, the first thing you should look at is putting a few more pounds of pressure in your tyres, which will open the treads out a little. You also need to bear in mind that because the grip will be lower, so will the weight transfer. This means that as you come into a corner there will not be so much weight going on to the nose and it will be very easy to lock up the fronts, which is why those with adjustable brake bias – a feature of race cars and specials – will wind it towards the rear in the wet.

Because there's less weight transfer in the corners in the wet you might also soften the anti-roll bars or the dampers, if that's possible with your car, which should make the car more pliant and give you a better feel for what it is doing.

But whatever you do to the car in the wet, the important thing is to try to enjoy the experience. There's no use letting a little water spoil your day, so just drive within your limits, keep your distance, and relax. Chances are you'll learn as much about yourself and your car on a wet track day as you would in ten dry days – and you'll have a ball in the process.

ABOVE Visibility can be a problem when it's very wet, especially if your windscreen's prone to misting, so think twice before binning that demister if you're lightening your track car.

LEFT The most important thing to remember when the track is wet? Enjoy yourself!

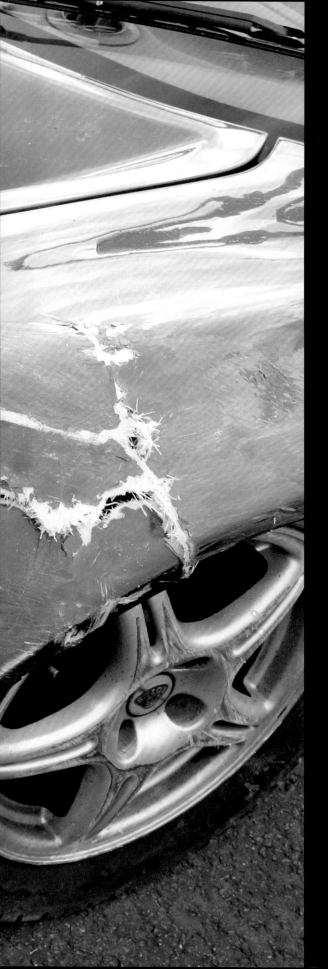

Insurance and recovery

Insurance and recovery

We've spent a great deal of time in this book explaining that track days are very safe, and if you've taken on board some of the advice in the preceding chapters then you should find they're even safer. But there's no getting away from the fact that crashes can happen, and even more common than accidents are mechanical failures. So what then?

Track days have a real aural appeal. There's nothing quite like the sound of hard-pressed engines or the squeal of rubber at its limit of adhesion. But there's one sound that's sure to sour the mood of any day. That's when the tyre screech is finished off with the sickly thump of metal on barrier and the cymbal-like smash of broken glass.

Oops! So what now? Well, crying is a good start, but how many tears you shed might well depend on whether you took out track day insurance or not before your day. Insurance is not compulsory on track days and many track day drivers simply don't bother. They'll argue that track days are very safe, given that you can control your pace and drive within your capabilities. And they may say that there's probably more chance of crashing on the road than on a track day. Which is a fair point, as on the road you're often at the mercy of maniacs and morons.

The sad thing is, though, that we all get carried away a bit at times, especially when the adrenalin's flowing, and let's not forget that part of the thrill of track driving is pushing that envelope just a little bit, isn't it? Also, there's always the problem of other people, and sadly there's absolutely no guarantee you won't get caught in another driver's moment. It's unlikely, but you still might want to consider a little insurance, and in the end that can only be your decision.

Take cover

When it comes to track day insurance there's an old cliché, one that is beloved by those that sell the stuff, and that is: 'Can you really afford to crash your car?' This, of course, is aimed mainly at those who are using their daily driver as a track day car. It's a fair question, and if your answer to it is a 'no' then it's certainly worth having a look at what sort of insurance package you can arrange.

But before you contact a specialist track day insurance company you really should check to see if you're already covered for track days through your regular road insurance. This is getting to be more and more unusual these days, but it's still worth checking, especially if you've held the same policy for a number of years. If you're lucky and this is the case, then it would be a smart move to get it confirmed in writing.

It's not just old policies, though. Recently at least one UK company has started to offer free track day insurance with its road car policies, although this is restricted to cars valued under £25,000, and there are others out there who will do something similar. You might have to pay more, and there will be a big excess, and perhaps even a restriction on the amount of days you do, but it's certainly worth shopping around.

All that said, the majority of insurance cover at track days still seems to be handled by specialist companies that started out in the related area of motor sport insurance. These brokers will provide cover for single track days or a package of days,

BELOW You'll usually find that track day insurance is cheaper for airfield days, mainly because there's little to hit and most incidents will result in harmless spins.

ABOVE Even at an airfield you can do some damage – this Escort looks a little the worse for wear after a trip into the 'rough'.

and they tend to have an in-depth knowledge of the scene, and hence a thorough appreciation of the risks involved.

All track day insurance companies will do things slightly differently, and one might suit you more than another, so it's always worth having a look at just what's available before you sign up. However, there's one point at which they all tend to converge, and that's on what they will actually fork out for.

Damage limitations

Most track day insurance companies will only pay out on the crash damage. It's unlikely that you'll find

a company that will pay out on damage to an engine or gearbox – although some might if it's proved either was broken as a direct result of the impact.

Another thing you'll find is that the excess tends to be, well, a bit excessive. Basically, the excess is the amount of money you'll have to pay out yourself whatever the full cost of the accident. In a way it's the amount you could reasonably afford to splash out yourself. So, if the damage resulting from a prang is beneath the excess, then there's no need to trouble the insurers. On the other hand, if it's a big one – the sort of shunt you've taken out the insurance for in the first place – then it's time to give them a ring. In this scenario, if you've done – let's say – £10,000 of damage, and your excess is £2,000, then you'll get £8,000 from the insurers to pay for the repairs to your car.

Incidentally, there's no third party insurance for track days, and this has caused problems in the past, most notably in the UK when a car being driven by a driver who was under tuition collided with another track day driver who then sued for damages. The resulting case was seen as something of a test case for the track day industry, and the final verdict was that anyone taking part on a track day is well aware of the potential risks involved and has the choice of *not* driving on the circuit. In other words, should anyone crash into you on a circuit, even when it's clearly their fault, then that's just bad luck… Which actually seems to be another good argument for taking out track day insurance, doesn't it?

RIGHT Bedford Autodrome has plenty of run-off, and because of this you'll often be given favourable insurance quotes for days that take place there.

Cover the cost

Premiums aren't cheap; indeed quite often they can come in at something close to the cost of the track day itself. But, in fairness to the brokers, that's to be expected given the nature of the activity. For some idea of what to expect, at the time of writing, for a Mk1 Mazda MX-5 worth about £2,500 for an airfield day, you would be looking at something like £60 – although some brokers will ask for a minimum of £80 – with an excess of £250. Then again, a £12,000 Elise will set you back £135 with an excess of £1,100 on the very same day – while on a permanent track you'll be looking at an extra 30 per cent or so on top of those premiums. When it comes to exotica, though, the prices will really make your eyes water, and a Porsche may well be charged around £300 for an event – but then the brokers say that nothing worries insurance companies quite as much as Porsches, Ferraris and the like, partly because of the cost of fixing them, and partly because of the speed they're going when they hit the barriers!

Guess you noticed the disparity between the premiums on the airfield and the circuit? There's good reason for this. Track day insurance companies need to assess the risk, and there's obviously far less to hit at an airfield – or the wide open spaces of Bedford Autodrome – than a regular circuit.

Similarly, a good insurance broker will almost certainly give you a better deal if your day is with one of the more professional operators, because they will know that accidents are far less likely to happen on a well-organised day. Which means you can get a good idea of just how good, and certainly how experienced, a track day operator is by the insurance quote you're given. This can actually be a good barometer of the quality of a track day.

It goes without saying that you should always check the small print when it comes to anything to do with insurance, and try to think of every eventuality on the day, too. For instance, if there's a chance that you might give someone else a drive in your car and they're rude enough to crash it, then there's a real possibility your insurance might not pay up – and this has happened – so check to see if there's an option for additional drivers if you think someone else might have a few laps in your pride and joy.

ABOVE The vast majority of track day incidents end up with a harmless trip across the grass, but many will still prepare for the worst-case scenarios by taking out track day insurance.

Claim game

If you're unlucky enough to have a crash at a track day, yet prudent enough to have taken out some insurance, then what happens next? Well, first of all – after you've sat there replaying the shunt in your mind like Steve McQueen in *Le Mans* – you need to get as many pictures of the damage as you can, and from every conceivable angle. These days this isn't so much of a problem, chances are you can take a few snaps with your phone, or failing that there's nearly always a photographer on hand who will be taking pictures of the cars out on track to sell to the punters as souvenirs, so try to get hold of him or her.

It's also worth getting some sort of signed verification from the track day operator that the accident actually took place on the day, and some of the insurance companies will have a form for operators and officials to fill out.

After the photos have been taken and the forms have been filled, then you need to inform the insurance company. You need to do it sooner rather than later, too – in fact one company stipulates it should be within 48 hours of the accident. You should also inform them if it's just a *potential* claim, as well. There's no use ringing them two months down the line when you've realised the damage is a lot worse than you at first thought. So if there's any possibility you might claim, then be sure to let the insurance company know as soon as possible.

One thing you should definitely not do, however, is make a false claim. Now, this cuts both ways; first you might think it's a great idea to pretend a car you've crashed on the track was actually crashed on the road. Well don't. The big insurance companies are wise to it. Some even employ track insurance experts to show them exactly how you can tell a car has been crashed on a circuit rather than the road – apparently it's very easy to spot. It's also called fraud, and people have been convicted for this very scam.

Similarly, if someone tries to pretend they've crashed at the track when in reality their mishap was on the road, then the track day insurance companies are also wise to it. Anyway, in these cases it usually only takes a quick call to the circuit or the operator, as track day crashes are rare enough not to be forgotten.

Life and limb

Personal injury insurance was once actually sold at track days in the UK, but since the Financial Services Act that's no longer the case. So if you want personal cover you had best make sure you sort it before you go – some track day insurance companies will actually sell it as an add-on to the car insurance. In other countries it will differ, so if it's a worry then check with the operator or your insurer.

That said, because of the peculiar nature of track days, there's a good chance that you're covered anyway by your regular personal injury insurance, if you have some. This is because a track day is a leisure activity and not a sporting activity, but you do need to check this with the insurers first, and make sure you go through the small print thoroughly.

One other thing worth mentioning is insurance for the Nürburgring. For proper track days at the 'Ring it's very expensive indeed – if you've been you'll know why – but for the public days it's all a bit of a mystery. Some say it's a public toll road, so your insurer will have to pay up, but the real truth of the matter seems to be that most insurance companies have cottoned on to the fact that here is a place with some very solid, and seemingly magnetic, scenery and they have taken care to put something in the exclusions about 'derestricted toll roads' and the like.

Either way, whatever the case with your own insurer, you can be certain there's no way it will pay out for those peculiar 'Ring bills for damaged Armco and loss of track time to other users. But, as always, the best course of action is to check your insurance company's position first.

Recovery position

You might be surprised to learn that at least one of the top UK recovery companies will pick up cars that have broken down – though not crashed – at a track day. At the time of writing the AA's Relay policy clearly states under the heading of 'General *Exclusions* from Service' [its emphasis]: '...vehicles broken down as a result of taking part in any "Motor Sport Event", including, without limitation, racing, rallying, trials or time-trials or auto test. However, for the avoidance of doubt, the AA does not consider concours d'elegance events, *track test days for road legal vehicles,* or rallies held exclusively on open public highways where participants are required to comply with the normal rules of the road, to be 'Motor Sport Events'. Seems clear then, if you have AA Relay and you break down on a track day in your road car, then that 'very nice man' should come and pick you up.

As for the other ones, a few phone calls seemed to suggest the situation was unclear, so best check your paperwork thoroughly before heading off, and if there's nothing to say you can't be recovered from a 'non-competitive' event, then make sure you've got all the documentation with you to argue your case. Some have even been known to push their cars out on to the main road if they're in doubt, but that's just a little bit naughty.

Whatever, if you've crashed – the main recovery companies only cover breakdowns, not accidents – or if your recovery insurance doesn't cover track days, then you'll simply need to arrange your own recovery, which can be very expensive indeed. Otherwise, you'll need to find someone with a trailer to help you out.

Another option is to take out some vehicle recovery insurance from a track day insurance company. At the time of writing one of the specialist track day insurance companies in the UK is providing this and it does seem like a good idea. It's cheap, at around the £25 mark, and it will cover your recovery needs for the first 70 miles from the circuit. After that it's £1.90 a mile. Clearly it's great if you're at your local track, but it could work out to be expensive if you're travelling far and wide for your track day thrills.

Of course, if you drive within your limits, and maintain your car, and don't over-stretch it, you needn't worry too much about any of this, but there's nothing wrong with planning ahead – just in case.

ABOVE If you trailer your car to track days, then recovery after a shunt is not a problem – but if you haven't a trailer it can prove expensive.

BELOW If you break down on the circuit you will be rescued, but if you haven't got a trailer you might need to think about how you are going to get the car home. (*Bresmedia*)

Safety kit

Safety kit

Track days have a fantastic safety record, but if you want to make sure you're just a little bit safer then there are a number of things you can do – both in terms of what you wear on the day and the modifications you can make to your car.

OPENING SPREAD Luckily accidents on the scale of this racing shunt are rare at track days – but if you do want to make your car a little safer for circuit use there are a number of modifications you can consider. *(Jakob Ebrey)*

There's an old saying in racing, and it holds true for track days: 'If you've got a 10-dollar head, then buy a 10-dollar helmet.' The point is, you really shouldn't scrimp when it comes to your headgear. Come to think of it, that saying is probably more relevant to track days than it is to racing anyway, because in the latter there are strict standards to adhere to before you're allowed to race, while for a track day the standard and quality of your helmet is pretty much down to you. So, on your own head be it – so to speak.

You'll almost certainly be obliged to wear a helmet on most track days – although it seems this isn't always the case in continental Europe. Even if you find an operator that will allow drivers on circuit without helmets, wear one anyway. It will make a massive difference should you go off track, and there's nothing like wearing a crash hat to put you in the mood.

Dress code

It should go without saying that you really need to choose your helmet carefully. While any helmet, within reason (you can't turn up with something that was designed for skateboarding or one last used in the Second World War) is OK to use – and good motorcycle lids will do the job – those designed for the race track are ideal. Modern race helmets are very strong and incredibly light, so if you're serious about your track days then get along to a racewear supplier and invest some time in choosing wisely. Make sure it fits well and is comfortable, and take some advice from those at the shop; they will know exactly how a helmet should fit.

One thing you should think twice about is buying second-hand, for there's no way you'll know if it's ever been dropped, and that's important. Helmets are designed to give a bit during an impact, so if it's been dropped the structure could well be damaged, and that will not be visible to the naked eye.

With that in mind you really need to look after your lid, and there are plenty of good helmet bags and cases on the market which could be worth investing in. If you do drop it, though, then you'll simply have to replace it. You just cannot take a risk when it comes to your helmet.

Incidentally, if you're not sure if you want to invest in an expensive helmet before you've tried a track day to see if you like it first – you will! – then you'll be glad to know that many of the better track day operators have a stock of helmets to hire out, usually for a very modest fee.

Whether you go for open-face or full-face is pretty much up to you (many find the latter claustrophobic in a closed car) but be aware that some track day operators have been known to insist on full-face lids if you're driving an open car. Personally, I'd go for full-face anyway, as these days they're very light and comfortable, and there have been some very nasty accidents to touring car racers and rally drivers over the years involving open face helmets.

Whichever you opt for, another item of headgear you might want to look at is a balaclava. Fire retardant balaclavas are a big part of the race driver's ensemble, and while there's little risk of fire in a road car they do stop the inside of your helmet from getting soaked in sweat – you'll be

FAR LEFT A good quality crash helmet should be a top priority.

LEFT Most of the better track day operators will have a stock of helmets to hire out to those who have not got lids of their own.

surprised just how hard you work on a track day.

But, however hot the work is, one thing that's a definite no-no on the majority of track days is short sleeves and shorts. Now this might seem a bit over-zealous on a summer's day, but there's a very good reason for the rule. A large part of the surface area of a road car is made up of glass, so when it goes off and hits something solid you can bet there's a lot of broken glass flying around, and you'll want to make sure as much of your skin as possible is covered. Long-sleeved T-shirts are useful, especially if your car gets hot inside, but otherwise a light fleece or a jumper will do.

Don't forget your hands, either. A good pair of driving gloves, while rarely mandatory, are a wise investment. This is not just because of the protection they give; driving gloves will also stop your hands from slipping when your palms begin to sweat – especially if your car's fitted with one of those horrible plastic steering wheels.

Proper race driving boots are worth thinking about, too. These thin- and flat-soled items of footwear not only look cool – if you like that sort of thing – but they also tend to be light in weight and they give you a much better feel for the pedals.

If full-on race boots seem a bit over the top, then have a look at the ranges of 'leisure shoes' many of the racewear manufacturers are now bringing out. These tend to be based on the full race boot and have flat and thin soles, so they're just the thing for track driving and you could even wear them day to day.

Even if you don't splash out on a pair of full race disco boots or track shoes, you might want to think about the shoes you're wearing, and certainly avoid those with thick soles, which kill your feel for the pedals, or those with overhanging uppers, which can easily snag on the wrong pedal at the wrong time. Not good.

Some track day regulars don't stop at pukka race-style gloves and boots, and it's not unusual

ABOVE The author in romper suit – at most days you won't look out of place in the full racing driver kit, and I tend to wear it if I'm driving something that's race-spec.

Race boots are not just for posing, they are also lightweight and they will give you a much better feel for the pedals. (*Demon Tweeks*)

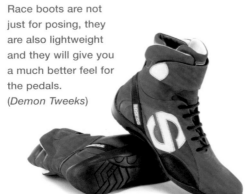

If race boots seem a bit over the top, then why not invest in some of the leisure shoes based on race boots that many of the racewear companies are now producing? (*Demon Tweeks*)

to see drivers fully togged up in Nomex romper suits. It's a personal choice, really, and you certainly won't look a wally if you're wearing all the kit – especially if you're driving a no-holds-barred special, but be warned that this gear can be quite expensive.

Many even say that wearing the full racing driver kit helps them drive better: look like a racing driver, drive like a racing driver, they say, but I'm not sure about that. Still, each to his own, and while I tend to wear regular clothes in a saloon I will often haul on the Nomex if I'm doing a day in something like a Caterham. It just feels right.

If you're going to wear racing overalls, though, it's important you keep them clean. You wouldn't believe how many drivers will work on their cars while still wearing their overalls, getting oil and even petrol on them, which sort of defeats the object when it comes to fire-retardant suits.

One more thing about suits; try not to choose one which is too tight. You really would be better off going a little bit on the baggy side, maybe not as far as Jacques Villeneuve goes, but a little loose all the same.

In fact, while trying on your suit at the racewear shop you should make sure you're comfortable in the position you're going to be driving, so if there's a race seat handy – as there will be in most shops – then sit in it and go through the motions; but don't make the noises, unless you feel you have to.

Belt up

When it comes to making safety modifications to the car, then how much work you put into it depends largely on your car and whether it's a bespoke track car or your daily driver. This is because, ironically enough, many of the more obvious safety mods you see on full-blown race cars are not necessarily completely safe when driving on the road.

Take harnesses, for example. On the track a racing harness is a wonderful piece of kit, holding you tightly in the seat so you can really feel the road beneath you – especially when used in conjunction with a good, snug sports seat – while also restraining you in the event of a crash. On the road, though, these can be far from practical, particularly when you need to lean forward to get a good view of the traffic at junctions. So, if you do fit safety harnesses and you're using your road car as a track car, then maybe think about retaining the inertia-reel belts for use on the road, if this is possible.

It's also crucial you make sure you wear the harnesses correctly. It is important that the lap belt

runs over your hip bones and does not ride up to sit on the soft part of your belly. The problem is that when you tighten your shoulder straps this automatically happens, so you need to make sure the lap strap is very tight first of all and then do the shoulder belts. And when it comes to *tight*, that means very tight. There's a Formula 1 saying when it comes to tightening belts that should always be kept in mind: 'tighten it until it hurts, then one more pull.'

Of course, all this is only any help if you've fitted your belts correctly in the first place, and the best advice is to follow the manufacturer's fitting instructions to the letter. But the main things to

ABOVE Most track day operators will insist that your arms and legs are covered when you're a driver or a passenger on a track day.

bear in mind are that the shoulder belts are to hold you back in the seat while the lap straps are to pin you down. Incidentally, if you go the whole hog and fit a full harness – which is called a six-point rather than a four-point – you also get a crotch strap, which is to stop you 'submarining' under the lap strap and into the footwell in the event of a really big shunt.

In the UK, the Motor Sports Association's Yearbook – the *Blue Book* – is the Bible when it comes to all the rules and regs of British motor sport, and in this there are handy diagrams on how to fit harnesses, with the ideal being for the rear restraining straps – from the shoulder straps – to extend horizontally to the rear of the car, while the lap straps are on a vertical plane. By the way, if your car is a daily driver you can be sure this won't be popular with back seat passengers.

One way to get some of the advantages of a harness but without the hassle and compromise of fitting one is to invest in a small device known as a CG-Lock. This is not a safety device, but it does give proven benefits when it comes to stability and control. The idea is it tightens the lap strap around your hips, keeping you pinned into the seat, which stops you slipping around in the seat and means you do not have to brace yourself under braking or cornering. It gives you more feel for what the car is doing beneath you and is a real benefit out

FAR LEFT If you're using a harness, make sure it's done up tight.

LEFT Fitting a CG-Lock to your regular belts will give real benefits out on track. (*Bresmedia*)

on track. In fact, in terms of speed gained over a lap it's probably one of the most cost effective little tweaks you can make. And you won't need to change your car to use it, as it's easy to fit to the regular belts, and it's easy to take it off, too – although it is actually as much use on the road as it is on the track.

Getting cagey

Fitting a roll cage has obvious advantages from a safety point of view, but also has the benefit of stiffening up the bodyshell of the car, which many think is worth the extra weight involved. Yet, while for a full-on track car a cage is an essential, when

LEFT A full race car safety set-up – as seen with this VW Racing Cup Polo – is obviously the best bet if you've a car that is purely for track use. (*VW Racing*)

ABOVE Roll bars make good sense in an open-topped car.

ABOVE It's a sensible move to carry a fire extinguisher in your car – just make sure it's securely fitted. (*Demon Tweeks*)

it comes to daily drivers that double up as track cars it's not so clear cut.

This is because on the track you're wearing a helmet, but on the road you aren't, and when your head hits a roll cage bar it will hurt – that's the case even if you just bump it when climbing in, never mind what might happen if you were to crash. Of course, you can pad the cage, but that's not really going to help in the event of a big one. Indeed, it would be fair to say that, in the case of roll cages at least, making a car safer for the track could possibly make it less safe for the road. And less practical, for a cage will take up much of the space where the back seat passengers would have been, and it's also going to mean it's much more difficult to clamber in and out because of the side-impact bars most good cages have. Fitting a cage is, then, a major step down the path to owning a full-on track car. That said, there are bolt-in cages on the market, and if you can be doing with the hassle of fitting it before every track day, this could be the solution.

If you have a convertible, then a roll bar or hoops of some kind are obviously a very worthwhile

investment. Be careful, though, some – but not all – of the Audi TT style hoops you see for small roadsters are little more than style bars, and the same is even true of some of the roll cages on the market. The best advice is to go for something that's been designed for competition, like an FIA-spec cage, because then it really should be up to the job.

While we're on the subject of keeping a balance between road and track use when it comes to the modifications you make, you really should be careful about stripping out the car to take weight from it if you're not going to go the whole hog and turn it into a fully kitted out track car. Apparently there have been some nasty, and wholly avoidable, accidents resulting from the sharp and unprotected edges that this sort of 'chuck it all out' approach can leave in the car.

Burning issues

According to Hollywood you just need to slam the door of a car too hard and it will burst into flames, yet it's actually very rare for a car to catch fire after a crash on the race track. That said, fires do happen, although usually as a result of a loose fuel

pipe or some other mechanical oversight rather than an impact.

There was a time when some track day operators would insist you carried a fire extinguisher in the car, and even though that is rarely the case these days it is obviously a good idea to have one to hand. It need not be a major hassle to fit one, either – just make sure you can get at it easily and that it is mounted securely, because something like an extinguisher is not the sort of thing you want flying around in the event of an accident.

A fully plumbed-in fire extinguisher system, as found on race cars, is probably a bit excessive, but if you're going for the full track car effect then it's certainly worth thinking about. In the end it's your decision, and it's all about weighing up the risk.

Weighing up the risk

The great appeal of track days, to my mind at least, is the freedom they offer: freedom from traffic jams and speed cameras on the one hand, but also freedom from the strict regulations of motor racing on the other. The fact that you can go to a circuit in

a regular road car and drive it as fast as you like is quite amazing when you think about it.

The thing is, to do so is your *choice* (that's the very nature of freedom) and it's also your choice whether or not you rig your car with every motor sport safety feature you can find. Whether your car is a bespoke track car or not will have some bearing on your decision – as we have seen above – but also your own attitude to track days should be taken into account. If you know you'll only be happy driving at 10-10ths – the limit – for every lap and every corner, then maybe you should look at getting your car seriously kitted out – or you should start racing!

If, on the other hand, you're willing to moderate your pace, and realise there's no pressure to push on the really tricky bits, there's no reason why you cannot enjoy yourself thoroughly and safely in a car that's as much at home on the shopping run as it is on the track.

The only caveat to that is: never, ever accept second best when it comes to the quality of your helmet – unless you actually do happen to have a 10-dollar head.

ABOVE Fires are extremely rare at track days but, as this driver discovered at Cadwell Park, there's always a chance.

Brakes and tyres

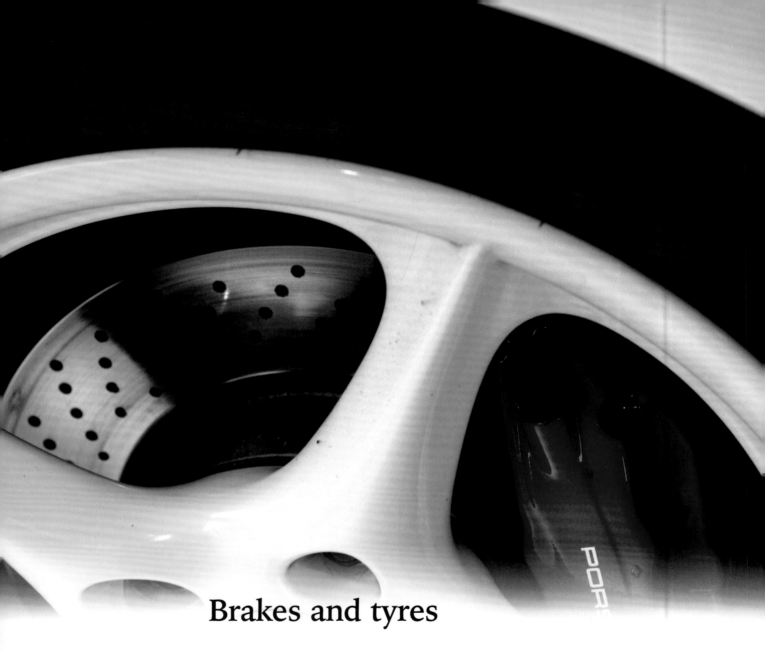

Brakes and tyres

Look after your brakes and there's a good chance they will look after you; and a beefy upgrade might be worth thinking about, too. As for the tyres, you should aim to check the pressures often, while also keeping an eye on the way they wear during the track day.

Brakes

It's one of the ironies of fast track driving that so much thought and talk is devoted to slowing down. Yet it's not so surprising, since brakes are the Achilles heel of a road car on the track. Indeed, it's often the poor performance of a car's brakes that will disappoint a first-time track day driver the most.

The problem is that brakes are asked to do far more on a track than they're ever asked to do on the road. In its day-to-day environment a car will rarely be braking heavily; in fact the only time it will be braked to the same degree as it is on a race track is probably when it needs to perform an emergency stop.

On the track you'll be bringing the car down from high speed to low speed a number of times a lap, and when there's a tight corner following a long straight it could mean one very big stop indeed. So you can be sure that just one fast lap of a race circuit will be a hell of a lot more punishing than your average drive to work. And it's all to do with heat...

You see, in essence brakes are simple enough devices. They apply friction to a wheel to slow it or stop it. The thing is, friction generates heat, and heat can generate its own set of problems. Or at least it can if it hasn't had the chance to dissipate, such as when the car is braking from high speed to low speed corner after corner, lap after lap, as it is on a track day. This means the brakes can quite easily overheat.

Fade away

The main symptom of overheating brakes is brake fade, which can either be the result of the pad overheating, of using new pads which have not been fully 'cured', or of the brake fluid actually boiling. In the first two cases, relating to the performance of the pads, it will usually result in a loss of the brakes' effectiveness rather than the spongy feel to the pedal which is characteristic of the third case – though other factors might mean you get sponginess as well.

A pad overheating will cause fade because

the binder resin in the pad actually melts, and lubrication of any sort between pad and disc is the last thing you want, for obvious reasons. When the brakes are fading because the pad is not fully cured, on the other hand, it's known as green fade. This occurs because some of the organic elements in a new pad can turn to gas when over-heated if the pad has not been fully 'cured' first. This is less common than it once was, mainly because most modern pads are pre-baked, but it's still worth going gently with a new set of pads to begin with to help them to cure. Anyway, it's always important to take it easy on a set of new pads to bed them in properly.

Brake fade will usually make its presence felt progressively. You'll find you need to make more of an effort on the pedal at a particular corner, or sometimes the pedal will feel spongy (see below). Both are signals that the pads or the fluid are over-heating and that the brakes need to be cooled.

These are not signals you should ignore. If you carry on regardless you could experience severe fade, which is pretty much like having no brakes at all, and clearly that's not an experience to savour. So, if you feel your brakes are overheating, slow down a bit and give them the chance to cool down – then they should gradually come back to you.

Even better, maybe you should pit. That's not to say you should dive into the pits at the first

FAR LEFT Porsche brakes tend to be better than most – which is one of the reasons 'Porkers' tend to be a popular choice for track days.

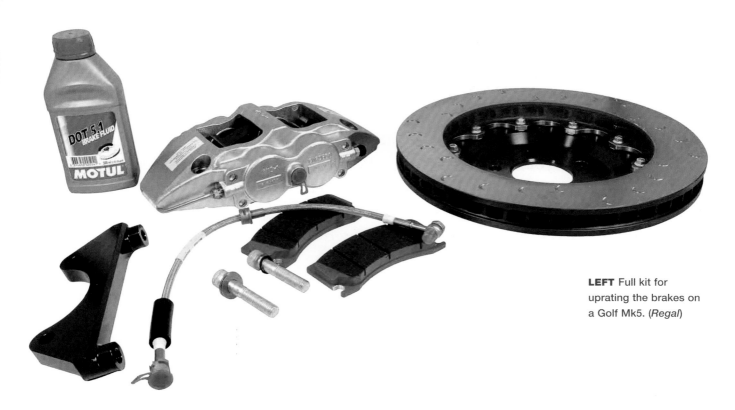

LEFT Full kit for uprating the brakes on a Golf Mk5. (*Regal*)

opportunity, though, as you really need to give your brakes a chance to cool first, and the best way to do that is by taking advantage of the air flowing around them out on the track. So, do a cool-down lap first (or even two cool-down laps), driving slowly, while making sure you keep out of the way of other cars, and using the brakes as little as possible.

This is very important as it helps to avoid heat soak in the braking system when the car is stopped – when the heat from the red-hot brakes is conducted to other components – which can cause all sorts of nasty situations, such as melting the grease in the wheel bearings or the pads actually catching fire.

Once you're back in the pits you need to be very careful not to rest your foot on the brake pedal, and you should *never* put the handbrake on – this applies even if your brakes haven't overheated, incidentally. This is because the red-hot surfaces of the discs and pads can fuse together, which in turn can cause pieces of pad to stick to the disc, resulting in brake judder – though it should soon wear away. More of a problem, however, is that it can also actually cause the disc to distort, which might mean you'll have to bin it.

It's common sense, of course, but don't forget to leave the car in gear (or even wedge a chock under a wheel) when you're in the pits with the handbrake off.

Boiling point

Sometimes you might find that the pedal will feel spongy, even to the extent that you can press it all the way to the floor – not good! This is often the result of boiling brake fluid. Brake fluid is hygroscopic (which means it can absorb water from the air very easily) and if water seeps into the system it will eventually drop the boiling point of the fluid until the heat generated by the brakes boils it, forming gas bubbles in the system that are easily compressed when the brakes are applied. So when you press the pedal you're compressing the gas rather than pressing the pads against the discs.

If this gets really bad, as if there's just air under the pedal, then it might mean you'll need to bleed the brakes, so make sure you have some extra brake fluid to hand – and keep it sealed so that the air can't get in. But generally, if you go easy on the brakes and you can live with a little bit of sponginess in the pedal, it's not too much of a problem.

You might consider bleeding the system after every track day if it proves a real headache, and even if doesn't it's probably worth bleeding

it fairly regularly if you're doing regular track days, just to make sure the boiling point of the fluid is not slowly going down as moisture gets into the system.

A better solution is to change your brake fluid to a type that's more suitable for track use. In the UK fluids are rated with a DOT (Department of Transport) number, which is to do with their chemical composition and boiling points. DOT 5.1 is pretty good for track days, but there are also some DOT 4s that will do the job admirably. These are all glycol-based fluids, which are by far the most popular – and hence easier to get hold of in an emergency – but there are also silicon-based fluids (DOT 5), which give high temperature performance while not absorbing water. They have their downsides, though, as they're more compressible than mainstream fluids, they're expensive, and the fact that they don't absorb water means that moisture can sit in the system and can cause corrosion. Also, bear in mind that DOT 5, unlike the other DOTs, cannot be mixed with other fluids.

Knock it off

A couple of years ago I was lucky enough to blag a race in a superb little hot hatch. It was huge fun, except that I would arrive at one particular corner with no brakes lap after lap. By the next corner it was OK again, only for the pedal to go to the floor the next time I arrived at that very same corner.

It was baffling, but some time after the event a race engineer I was talking to explained that I'd probably experienced something called brake pad knock-off. The cause turned out to be a chicane before the corner in question which had some very rough kerbs that you just had to use if you wanted to be quick.

The problem with this, in some cars at least, is that the discs can wobble when the car hits a kerb, and this little shimmy can force the pads back into the calipers so that they aren't in their usual position – which is almost, or just, skimming the disc – when you get to the next corner. This then means that the pedal travel you would normally use for braking is used up taking up the slack (for want of a better word) and you go into the next corner experiencing something akin to severe brake fade, or even brake failure. Let go of the brake and then hit it again and the problem will usually have rectified itself, but by that time you could already be getting closely acquainted with the scenery.

Even worse is that the natural reaction when you realise you've no brakes is to brake harder, and as all the braking is now on the rears this tends to lock them and throw the car into a spin.

All this is another reason why it's worth giving yourself longer braking areas on a track day than you would during a race, and you would also do well to avoid the bigger kerbs and rumble strips to stop the knock-off in the first place.

Upgrades

When it comes to upgrading your car for the track, the brakes are definitely the right place to start. In fact, you'll probably be looking at fitting at the very least fast-road pads quite soon after you've completed your first track day. That said, some light cars can certainly get away with using their regular pads, as long as you look after them and don't expect to do too many laps at a time.

If you do think about uprating your brakes, though, make sure you talk to an expert first, and preferably someone with knowledge of your car. Also, think about getting that expert to do the work, too – for there can be no 'that's close enough' when it comes to your brakes.

As with most car upgrades for the track, you'll find that one step invariably leads to another, and in the end you'll want to upgrade the entire system – but most people do start with a quite simple change of pads. There are plenty of performance pads on the market, too, ranging from fast road to full race versions, and what you choose will depend largely on whether you have a bespoke track car or if it doubles up as a daily driver.

If you have a track car and you trailer it to track days, then it goes without saying you should fit the best race pads you can get – just be sure to warm them up thoroughly before you use them in earnest, and bear in mind that new race pads will take longer to bed in than road pads. But if yours is a daily driver, then things are a little more complicated.

You see, race pads – sometimes called 'hard' pads as opposed to 'soft' road pads – need to be up to temperature to work efficiently, and you're unlikely to get them up to temperature on a run down to the shops. This is where fast road pads – which are neither too hard for the road nor too soft for the track – might be worth looking at, as they're the perfect compromise in many ways. There are also plenty of pads on the market that are designed specifically for track days which might be worth looking at, too.

Some drivers even change their pads just for the track day, which does make sense if you've a car that's so heavy on the brakes you really do need race pads. It can also be worth taking along a set of spare pads anyway, as you really don't want to drive home without any meat on your pads. This is true even if you've done a day before without any problems, as all tracks are different and some circuits will take a lot more out of the anchors than others.

Because most of the braking effort is at the front (around 80 per cent of it in fact) many drivers will only upgrade the pads at the sharp end on their track cars, but you should perhaps think twice about this as it means the rear pads could be coming up to temperature first, which might cause rear wheel lock ups. Best to keep the braking system balanced then, fitting the same pads fore and aft.

There have actually been great strides in brake pad design over the past few years, with composite pads, such as carbon metallic, becoming popular. Some people love them, some are not so sure, but they certainly last a long time and – in my experience at least – seem to give good pedal feel. But that's the thing with brake pads, what you'll like will depend very much on how you drive, and a hard race pad needing lots of pedal pressure will certainly not suit every track day driver.

If you find an upgrade in pads hasn't stopped your brakes from overheating, then you might want to look at cooling them more efficiently with the use of strategically placed air scoops or hoses directed at the front discs; but you do need to be very careful how you route these as they could snag on suspension or steering components.

As for those discs, the easy answer is to go for as big as you can get – the greater the area, the more heat you can dissipate, but much more important is that the braking system will be more efficient because of the better mechanical leverage a bigger disc provides.

Replacement hoses can be a useful way to upgrade your braking, too. Regular rubber hoses can expand with the massive braking involved in track driving, which means some of the braking effort does not find its way to the pad – another reason for a spongy feeling in the pedal. The simple solution is to fit sexy-looking steel braided hoses, which have the added bonus of being far more resistant to damage.

Warning

There's every chance that a brake upgrade will be the first thing you attempt when it comes to making your car better suited to the track. As far as pads are concerned you should always bear in mind how you use your car – full race pads are next to useless on the street – and keep an eye on how they wear.

As for the rest of the braking system – the shiny stuff like discs and calipers that always look great in the product pages of motoring mags – try to go for the well-known, reputable brands. It will cost you, but when it comes to your brakes you do not want to cut corners.

Also, when it's *anything* to do with brakes, don't be afraid to consult an expert, and make sure it's a track preparation expert rather than your local bolt-and-go tyre and exhaust centre. That's important.

Tyres

The wheel deal

Your tyres are your point of contact with the road; all of your communication with the track surface up to the goodbye of a spin-off takes place through the medium of the rubber. They're pretty vital then.

But before we get to the tyres, what about the wheels? These days most performance cars come with flash alloys, a set of which are often worth about as much as a cheap track day car in themselves. But track work can spell the ruin of shiny rims, which can get coated with extremely hot brake dust, while the heat from the brakes can also discolour your alloys – all of which shouldn't really worry anyone with a true track heart, but then again it could hit the resale value of your car. A spare set of wheels just for track days could be

the answer, and this might also sort out another potential problem in the process, getting the road-track balance right when it comes to the tyres.

Tyre choice

If you have a track car that you intend to trailer to track days, then the choice might seem simple – just bolt a set of slicks on to it and off you go – and the same may be true of a car you drive to the track with a set of spares stuffed into the back. But a word of warning here: slicks (treadless race tyres that are only suitable for dry tracks) can be difficult if you're a beginner. Sometimes they can break away suddenly rather than progressively and they certainly need to be warmed up thoroughly – particularly the rears on a front-wheel-drive car, something which is a bit of an art in itself.

Also, by fitting slicks you'll be putting far more severe forces through the suspension – lateral loads the car will not have been designed for. There's even the chance you could suffer from oil surge if you haven't fitted a baffled sump to combat the increased cornering forces, of which more in the next chapter. And then there's the

BELOW Some drivers will come to track days with a choice of slick or wet tyres, but there's no reason why you can't lap quite happily on your regular road tyres.

RIGHT There's some sexy rubber on the market these days – but you need to think about exactly what you want from your car before you decide what you should fit.

RIGHT At some of the better track days there will be tyre companies in attendance, and often there will be tyre-fitting facilities in the paddock.

need to carry another set of tyres in case it rains – though you'll have your road rubber to fall back on if it's a daily driver – and, perhaps most important, there's the fact that a few operators will simply not allow cars to run on slicks.

Clearly then, there's a lot of thinking to be done before you go down the route of fitting slicks, as there should be with any change of rubber. Here's a for instance. Most think that wider tyres will always offer better grip than narrow tyres. True, to a certain extent, but there's simply no point in running a light car with massively wide rubber, as there won't be the weight in the car to get all that rubber to work for you. A tyre needs to be able to deform in order to grip the road surface, and it can't do that if there's no weight acting on it – if this is the case the car is 'over-tyred'.

So think of it in terms of the car itself pushing the tyre into the road surface, and unless you're going to add wings and things – and then it gets very complicated – you would be well advised to be cautious when it comes to fitting over-wide tyres. Your best bet is to see what tyres others with the same car are running, and ask them how they're getting on with them.

These days there's actually some great track day rubber, that's also road legal, which is definitely worth a look at if you're going to get serious about your track driving. Some of these tyres have little in the way of tread and are getting close to slicks, though, and it has to be said that not all of these are as good in the wet as regular road tyres.

In fact, the best tyre choice for the wet would be a regular road tyre, the newer the better, with deep channels to disperse the water. But there can actually be a problem with new road tyres when the track is dry, and that's because the tall tread blocks will tend to move about a lot and overheat, which means you'll get a lot of wear out on track and rather less in the way of grip. You could even destroy a set of brand new tyres by over-driving on them when they're overheated.

With that in mind, if you're using regular road tyres for track days then it's worth making sure they have worn down a bit first. But in the real world that's not always an option, so you may just have to take it easy to begin with, wearing them down slowly and resisting the temptation to slide the car too much.

Of course, you should always keep an eye on the way your tyres wear throughout the day, as this will give you an idea of how the suspension is working for a start, and on whether you're running with the correct tyre pressures, too. Don't be surprised if it's the outer edge of the outside front

tyre, particularly on front-wheel-drives, that takes a lot of the punishment either – be especially careful to keep an eye on this. And remember, if it's your daily driver you're using you'll need your tyres to get you home – the police will not look kindly on

anyone driving on bald or badly worn tyres on the road, and the fine for this in the UK will make your eyes water.

One other thing, always make sure you inspect your tyres for damage *before* you leave for a track day. Look out for any damage to the treads or cracks in the sidewalls; the sort of faults that can cause problems when the tyre's under real stress on the track.

Under pressure

If you're serious about track days, then one of the first things you should get yourself is a tyre pressure gauge. It doesn't need to be madly expensive, or even completely accurate – the main thing is that it's consistent. Some people will use the gauges on the air pumps at petrol stations, but while they can be accurate (some, at least), you can hardly drag them along to the track day to check your pressures throughout the day – and that's important.

There's a lot of talk about tyre pressures at track days, not all of it sensible. But as a rule of thumb you won't go far wrong if you start with the manufacturer's recommended tyre pressures if you've a lightweight car. If your car is particularly heavy, though, or perhaps not an out-and-out performance car, then there's a fair chance the manufacturer will have allowed for a little sidewall flex in the tyre to help soften the ride on the road. In this case think about putting in a few more psi, but not much, maybe as little as 2psi and probably not more than 10psi.

In fact, the sidewall stiffness is very important, for the stiffer this is the more responsive the car will be, but you still need to keep that flat tyre print on the road – your point of contact with the track

– so there's clearly a balance to be struck, and too much tyre pressure will mean there's a slight crown on the tyre, maybe resulting in only the middle portion of the tread sitting on the track.

As mentioned, you should aim to check your tyre pressures throughout the day, and make sure you do this when they're either hot or cold. If you choose to do it when the tyres are hot, then you really need to check as soon as you come in, so it's probably easier to be consistent if you check them when they're cold.

Why is this so important? Well, tyre pressures will increase out on track because of the expansion of the air inside – they can go up as much as 10psi, in fact – which is why you need to think hard before putting more air in before you go out. Race teams get around this by using nitrogen or a special air that has the moisture, which is the agent for the expansion, taken out of it. But that's a bit too much effort for a track day, to be honest – this is supposed to be fun, not an expedition to Mars.

You should always inspect your tyres throughout the day too, to check for excessive wear, but also to get an idea if your pressures are correct. A clear indication of tyre pressures being too low, for example, is any part of the tyre's sidewall being scuffed; while if there are signs that the outside edges of the tyres are doing little work – lack of scuffing, little heat – then perhaps you need to come down a psi or two.

But one thing you should certainly make sure of is that if you do put any extra air in your tyres, or take any out, you'll also need to take it out or put it back in again for the journey home. So as well as investing in a tyre gauge, you might also want to think about getting yourself a tyre pump.

TOP LEFT Keep an eye on the wear at the outer shoulders of the tyres – particularly the front left on the clockwise circuits of the UK.

TOP RIGHT The pressure seems to be a bit low on this tyre. The picture gives you a good idea of the sort of stress your rubber will be under in a corner.

13

Engine and suspension

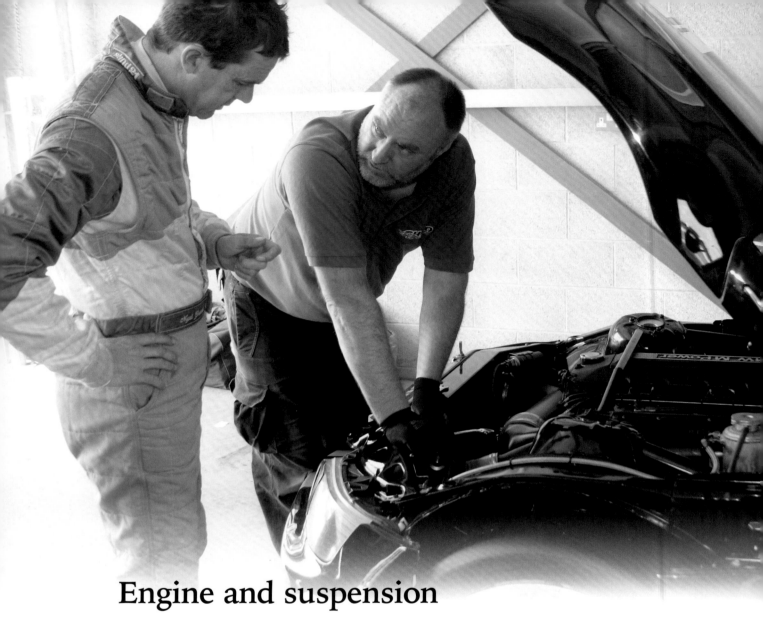

Engine and suspension

It's important to keep an eye on your engine's vital functions during a track day, while when it comes to modifying the motor or the suspension then bear in mind that anything you do is likely to have a knock-on effect elsewhere.

Unless you're in the habit of keeping your car in third gear on the motorway, it's doubtful you'll use as many revs on the road as you do at a track day, and since your engine will be working far harder on track you need to keep an eye on it. So while on the straights, make it a habit to check those gauges and warning lights – do it without thinking, the same as glancing in your rear-view mirror. And don't forget the temperature gauge. Overheating can be a problem at track days, especially with older cars. If your car does suffer from this, then a short-term fix is simply to use less revs, or you might even want to make your sessions shorter. As soon as the needle begins to creep above a certain point then use a little less revs on the straights and think about coming in, because lapping with an over-hot engine is just asking for trouble.

But don't pit straight away, you want to bring the temperature down slowly if you can, otherwise there's a chance that components in the engine might crack from contracting too quickly. Also, once you're in the pits you might want to leave the bonnet open, just to dissipate the heat, while those with turbochargers should keep the engine running so the intercooler can continue to cool the turbo.

Of course, it goes without saying that you should also check the water level throughout the day, but if you've an old-style rad' that you need to take the cap off for this purpose, then make sure the car has cooled before you do so.

As far as the cooling system itself is concerned, if you're looking to make it more efficient it's worth flushing it out and refilling it with fresh coolant. Another easy tweak would be to use one of the readily available additives that bring the temperature of the coolant down. But some cars tend to brew up more often than a gang of builders, and if that's the case with yours then ultimately you might want to talk to a marque specialist about how to go about uprating the cooling system.

It's not all about keeping cool, though, and you'll also need to make sure the engine is *up* to temperature before you start your fast lapping.

This is very important, as the oil will simply not be able to work properly until it is at the correct temperature.

On the subject of oil, use the best you can find and change it often is the best advice. Some will even change their oil before each track day they take part in, which is probably a bit excessive but it's good practice if you're doing a lot of days, while it's also well worth taking a bottle of oil along with you to keep the level up. Oh, and be sure that the oil cap is nice and tight, as these can blow off when the oil pressure builds up.

These days many people will use synthetic oil, but be warned that a fully synthetic lubricant could be a little too 'thin' for older engines which might be that little bit 'looser' than newer ones.

As for the oil level, you should be very careful not to overfill the engine for obvious reasons, but you should also try to fill it to the maximum. This is because the oil in the sump will be slopping around a lot when you're cornering hard, particularly in long curves, and it can 'surge' away from the pick-up pump and starve the engine of its lifeblood. Indeed, sometimes you might even see the oil pressure warning light flash in this sort of corner – not a welcome distraction in a fast turn when you're fully concentrating on staying on the road.

FAR LEFT Opening the bonnet once you've come into the pits will help to dissipate the heat.

BELOW Check the oil level regularly. If your car hasn't got a baffled sump then it might be worth filling it to the max to avoid oil surge – but be careful not to overfill it.

Put a newer engine in a lightweight old car and it will fly – this 16v Vauxhall lump was fitted in a MkII Escort – but before you try anything like this you should work on your driving and then see to the suspension. Bhp should be the lowest priority when it comes to track days.

Obviously, this is not a happy situation, and if it looks like it's going to cause you a real problem then you might want to look at fitting a baffled sump. There will probably be one available from a go-faster stockist if you're running a popular car, but it's also a job you can do yourself, if you've the skills, by simply welding or bolting metal plates into the sump. These little fences within the sump should help to keep the oil from surging.

As far as the fuel is concerned, while many cars will lap happily on regular unleaded petrol there are benefits in using higher octane super unleaded brews, and not just in terms of performance, as it also helps avoid pinking – predetonation in the cylinder – which can lead to all sorts of problems in the long term, such as overheating inside the engine.

One more thing, make sure you check all those vital belts in the engine before a track day, and bear in mind that your cambelt might have to be changed more often than it normally is – remember what we said about one track mile equalling 10 road miles?

Power priorities

Modifying your engine? Well, this is certainly the very last thing you should look at. Yet it's often the very first thing people *will* look at, mainly because many have a fixation with those three little letters *b, h* and *p*. But, in truth, bhp (brake horsepower) is pretty much just pub talk, a 'mine is bigger than yours' sort of thing, and when it comes to track

days it really should be at the very bottom of your list of priorities.

Why? Well, first there's undoubtedly far more performance to be found in your driving at the beginning of your track day career than in winding up the wick in the motor. In fact, it's been said that a couple of sessions with a good instructor beside you would equate to over £1,000 spent on the engine, if you were looking at lap times – which of course you're not on a track day, but I'm sure you get the point.

Also, there's no point in having big power if you can't use it, so you first need to build the platform on which it will operate – bigger brakes for coping with the higher straight line speeds and better tyres and suspension so that you can actually use that grunt and not spin it away in a cloud of expensive tyre smoke.

When you do get to the point where you're ready to modify your engine, though, then be methodical about it. A turbo might give you plenty of extra bhp, but will the bottom end be able to take the extra stress, and what about the extra cooling implications? When it comes to proper engine mods things get complicated.

This is especially so with modern engines, which are pretty much beyond the scope of most home mechanics these days anyway, thanks to all the complex electronic systems. Most engine tuning now takes place with a laptop at a specialist with all the necessary equipment on hand. And it's amazing how much power they can

LEFT If you're going to strip out your track car, then it might be worth going the whole hog and fitting a good cage, which will not only make it a safer place to be should the car roll, but will also stiffen the shell.

unleash with a simple change in the mapping – without even getting their hands dirty!

Lighten the load

Perhaps one of the most cost-effective, as well as performance-effective, modifications you can make to a car for the track is to strip out some of the excess weight. But you do have to be very careful here. If the car is not your daily driver – or you're not overly worried about creature comforts – this is certainly a cheap and cheerful way to get better acceleration, improved braking and a car that is kinder on the brakes while it will usually go round corners better, too, because you'll be cutting down the weight that will be transferred across the car.

The obvious things should go first: stereos, carpets, air con; and then it's on to the rear seats, maybe the passenger seat if you don't take passengers, the sound-proofing, and so on. Some even go to the extreme of replacing the glass with plastic windows.

But, as I've said, you do need to proceed with care. This kind of lightening tends to leave lots of sharp edges which can easy snag a flailing arm in an accident – and one track day operator tells me this is the most common injury he sees; in fact it's the only injury he sees! So if you're set on stripping out extra weight, then think about adding padding to the raw edges of steel, and maybe wear a race suit and gloves.

Ironically, those who go the lightweight route

soon end up piling the pounds back on again in the form of a roll cage. Apart from the protection this affords should you go over, it also stiffens the bodyshell, which in turn helps the suspension to work in the way it was designed to.

The same is true of strut braces, which are solid bars that link the top of the suspension turrets in the engine bay, though they can also be fitted in the rear of the car. Both will firm up the bodyshell in much the same way as a roll cage will – although there aren't the inconveniences associated with a roll cage, so they could be the ideal addition to a daily driver. Clearly, then, there are times when it's also worth putting on a little weight.

BELOW Strut braces are a great way to stop the bodyshell flexing around that rather large hole in which the engine sits.

ABOVE Don't let the body roll worry you too much, it can be all part of the fun – but don't be surprised if you get through plenty of front-left tyres. (*Steve Clark*)

Suspension

Many people are surprised by the amount their car will wallow and roll the first time they take it out on track. Even cars that seem to have firm suspensions on the road can feel all at sea out on the circuit, so when you do start to think about modifications then this will be one of the obvious places to start. But you'll need to be careful with the suspension. Change one thing and it will have repercussions elsewhere, while a full-on track set-up may not be great on the road.

Here's a good example of the sort of thing you might have to think about. A common problem with a road car on the track is that when the body rolls in a corner the suspension geometry changes so that the outside front wheel will cock itself into a positive camber – that is it will lean out from the top. This will mean there's not a good contact patch with the track, the tyre will overload and break into a slide more easily, and because of this you'll get nasty understeer which will then result in overheating on the outer edge of the tyre. You see this quite often, three of the tyres on a track car will be fine, while the tread on the outside front (usually the left front on clockwise UK circuits) will be worn away at the shoulder.

To counter this you may be able to dial some negative camber into the static suspension geometry – that's the way the car will sit when it's not moving – in the hope that this will give you a flatter contact patch on that front-left when the car's in a turn. That's fine if it's just a track car, but if you use the car on the road as well you might find the tyres will wear unevenly because of the way they will sit on the road in a straight line. And even on track, too much negative camber can lead to problems with stability at high speed and under heavy braking.

So maybe the best solution is to cut down on the amount the car rolls in the first place; perhaps

RIGHT In a corner the weight transfer on to the left-front is a limiting factor.

adding anti-roll bars, stiffening the springs, uprating the dampers, or lowering the car – or more likely some combination of all of these things. But this will also have an impact on your suspension geometry, so you need to be very careful about how you go about it, and talking to an expert is the obvious first step. Also, the ride on the road is bound to be harsher if you go for a stiffer set-up.

Even if you aren't too worried about that, you'll still need to compromise to some degree, as too stiff a set-up can lead to understeer, or to a car that has real problems under braking or when it's trying to put the power down, because it's simply skipping over the track surface.

But the real point here is that anything you do will affect something else, and there's no *one* answer to getting a car to work well on track. Everybody will have their own ideas, and often each is as valid as the next. One thing may well work very well on one car but not on another. And then there's the none-too-small matter of the personal preferences of the driver, which has as much to do with experience and driver skill – some just do not feel comfortable in an 'edgy' car – as it does with the car and what it's to be used for.

But before you start thinking about these compromises, think about why you're going on track days in the first place. If it's just for a little fun, then maybe you should just put up with the body roll and the tyre wear, and you should certainly think long and hard before investing in what can be expensive suspension upgrades.

It's worth checking over your suspension before you go on a track day, though, just to make sure there's nothing that's about to part company with the car; and while you're at it, also check the steering, looking out for any play in the rack, the column, and the steering arms.

If you do want to modify your suspension, then you could do a lot worse than invest in Alan Staniforth's superb book *Competition Car Suspension,* but in the meantime let's just go through some of the terms and the suspension components involved, and what they're supposed to do, so at least you can hope to speak the same language as those you might approach for advice.

Springs

Otherwise known as those bouncy things, but just how bouncy is the big question. You see, setting up your suspension is pretty much a compromise between keeping the wheels in contact with the track on the straights, and stopping the car from rolling too much in the corners.

You'll usually be looking at coil springs these days, and the object of the exercise is to find springs that are soft enough to be able to allow the car to ride over the bumps and the kerbs without it bouncing and losing traction, while being hard enough to stop the car from bottoming out under braking and cornering, and also to help quell the body roll as much as possible.

Most road car springs will be *very* soft, because even on performance cars manufacturers will want to give the driver a comfortable ride, and often coils with a much higher spring rate – that's the amount of force needed to squash it – will still be fine on the road and even better on the track, especially on many of today's smooth circuits. That said, you'll still need to have enough compliance to stop the car skipping over what imperfections in the surface there are, not to mention the kerbs, so rock solid may not be the way to go.

As a very general rule of thumb, the more grip you're getting from your tyres, the harder your springs need to be – to counter the roll from the extra cornering force generated by the tyres.

ABOVE Coil springs. (*Spax*)

Dampers

Often confusingly called 'shock absorbers', dampers actually control the springs rather than absorb the shocks. A spring, left to its own devices, will simply keep on going *boing-boing-boing* until it runs out of its, er… *boingyness*. But enough of the tech-speak, the point is a spring

ABOVE Dampers: sometimes confusingly referred to as 'shock absorbers'. (*Spax*)

then, they can be a driver's best friend.

Adjustable dampers are a great boon for track day drivers, by the way. They allow you to control the bump in cornering to give better control, if a harsher ride. Then you can reset them for better rebound and comfort for the journey home. Sorted!

Lowering

If you're fitting new springs, then it's probably worth thinking about lowering the car at the same time. This will bring the centre of gravity down and reduce weight transfer, while the car will almost certainly look a lot sexier to boot – which is a nice bonus.

The great thing is that most mainstream road car manufacturers will build in plenty of room for you to exploit, simply because a car, in its day-to-day life, must be able to shift four or five adults, or loads lashed to the roof rack, and all this sometimes on rough roads, without the tyres ever touching the inside of the wheel arches.

You need to be careful when lowering your car, though. Not just because of the obvious problems of the car bottoming out or the wheels fouling the inside of the wings when steering, etc., but because you can also upset the delicate balance between the car's roll centre (the imaginary point around which the car leans in a corner) and its centre of gravity, which you might think of as the point where you could theoretically balance the car.

You can also mess up the suspension geometry – particularly the camber – because the suspension arms will now move around differently, meaning that you'll have different camber changes in braking and cornering, and more problems with keeping those contact patches flat.

All this sort of stuff is designed into the car at the boffin stage, so you need to think long and hard before you mess about with it, and when you do it's best to approach lowering in increments. An even better approach would be to get professional advice.

Bushes

A relatively cheap modification, and also one which will have no real knock-on effects elsewhere, is to replace the suspension bushes. Manufacturers will often fit cars with rubber bushes, which not only tend to have a lot of play built into them but can also get sloppy over time, making for a 'soggy' response. The really good news is that the hi-tech replacement polyurethane or nylon bushes you can buy not only locate the suspension more accurately but also still keep the

ABOVE Classic 'outboard' suspension on this Caterham.

needs to be 'dampened', so that the car can return to its regular ride height as soon as possible after the spring has done its bit, without bouncing around too much. Dampers work in bump (which is compression) and rebound (which is extension).

Uprating dampers will probably make the biggest difference to the handling and cornering of your car. This is because dampers deal with what are called transient conditions, such as getting on the brakes and turning into corners – which is why race drivers obsess about them so much. Think of it this way: while springs – and anti-roll bars (see below) – are all about how much a car will roll, dampers control the speed of that roll. Clearly

noise down in the cab – one of the reasons for rubber bushes in the first place.

The next step from here would be to fit spherical bearings – rod-end joints – but then you really are going down the race car route and the ride will be a lot harsher, and noisier too.

Anti-roll bars

As the name suggests, the idea of anti-roll bars is to try to reduce the amount of lean in the corner. It actually resists the roll resistance on the front or the rear, thereby changing the handling characteristics – because of the way the weight is transferring across the car.

If your car hasn't got one to begin with they're not always easy to fit, but there are kits for many cars. They can be fitted on the front and back, though many road cars only have them on the front. Full-on race cars will tend to have bars on each end, and in this case they will tend to be adjustable, too. Adjusting the bars is a great way to help dial out handling imbalances, with a softer front bar – they tend to work by twisting, or flexing, 'hard' or 'soft' with degrees in between – and a stiffer rear bar helping quell understeer while the reverse should dial out oversteer. It's not quite that simple, but that's the general idea.

Alignment

A relatively cheap tweak, and one that is often overlooked, is to get the alignment sorted. That is to make sure the camber angles, caster angles and the toe-in is correctly set in the first place, or to check they're at their optimum settings after you've made a suspension modification. Get an expert on your make of car to do this for you – and not the local tyre fitter – and you'll be amazed at the difference it will make.

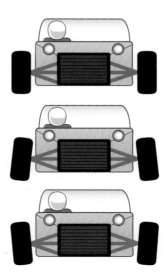

ABOVE If your track day car doubles as a daily driver, any suspension modifications you make are likely to be a compromise.

LEFT Fig. 15. From top to bottom: neutral, negative and positive camber.

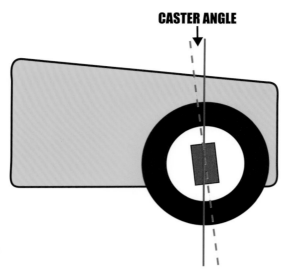

CASTER ANGLE

RIGHT Fig. 16. Caster will enable the steering to self-centre.

The language of suspension

The language of suspension is often made up of choice four-letter words, as fiddling with one thing usually messes up another, but there are some suspension words you can utter without upsetting the vicar.

Camber is certainly one word that will come up often in suspension discussions. It refers to the angle of the wheel – when looking at the car from directly in front or behind (Fig. 15). If the wheel is leaning in at the top it's said to have negative camber (think BMW), while if it's leaning out at the top it's said to have positive camber (think 2CV). Negative camber is also sometimes called knock-kneed, while positive is called bandy-legged, if it helps.

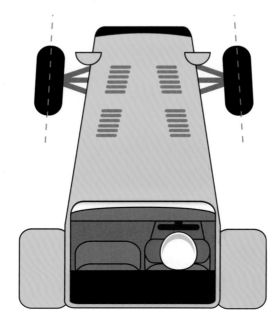

RIGHT Fig. 17. Toe-in is when the wheels are pointed inwards.

In a corner the outside wheel will usually take on some positive camber, because of the lean of the body acting on the suspension. Ideally you want to keep the contact patch of that wheel, and the others, as flat to the road as possible, so when it's possible drivers will often try to dial in some negative on their cars – especially if they're predominantly used on the track.

Caster, meanwhile, is an imaginary line that connects the upper and lower points of the suspension where they're attached to the front wheels (Fig. 16). The caster angle is always positive – sloping down towards the front of the car – and it is what gives the steering its ability to self-centre, while it also helps with straight-line stability. It's unusual to be able to adjust caster on a regular car, but if you've a special or a race car then be aware that lots of caster, though generally a good thing, can make the steering heavy, while it can also change the camber.

Toe refers to the angle of the wheels as seen from a plan elevation. 'Toe-in' is when the front of the wheel is pointing inwards (Fig. 17), while 'toe-out' is the opposite. If it can be adjusted, then a bit of toe-in can help a car turn in. However, note that toe-out at the rear will usually cause problems.

Something else that's a bit of a problem is *bump steer*, which is when wheels toe-in or toe-out because of the movement of the suspension, because it hits a bump or when the body is rolling – and this is not uncommon on road cars. It's well worth getting this sorted if you can.

Where next?

If you've got to the stage where you're mucking about with camber and caster, maybe even taking tyre temperatures with a pyrometer after lapping sessions to see how the suspension's performing, and you still feel you just want to go faster and faster still, then maybe it really is the time to move on up into the world of racing.

More and more people are coming to motor racing from track days now. Indeed, many will happily even take part in both, enjoying the amount of lapping on track days, and the competition at races.

Actually, if it's a bit of competition you crave, then there are a number of motorsport disciplines you can enjoy with your regular road car, such as autotests, autocross, production car trials, road rallies, and even speed disciplines like hillclimbing and sprinting (events which might be organised by your local motor club). More details can be obtained from the MSA in the UK (Appendix 2).

Many of those who get into motor sport via

track days do so because they have developed a track car so much that it is simply the next logical step. There's even one track day organiser in the UK (Lotus on Track) which has set up its own race series – The Elise Cup – such was the demand from its track day regulars to go racing.

Most others will have to fish around for a championship, though, which might not be easy to find, especially if you've modified your car so much it's simply too hot for a championship that might have seemed like a natural fit. In this case you may have to 'de-tune' it. It's likely, though, that you'll opt to trade it in for a car that's already race ready, or build one up yourself.

But before you can even think about going racing you'll need to get acquainted with a little bit of paperwork, and start to spend a bit of money. In the UK (some other countries are similar) you'll need to order a Go Racing pack from the MSA, and then go through a medical, which isn't too strict. After that you need to take your ARDS test at a racing drivers' school, which involves an on-track assessment and a written test.

Neither the assessment nor the test is difficult. If you're a track day regular and you've had a bit of tuition in your time you'll sail through the first, while the test is easy enough as long as you've done your preparation – which is outlined in a video you get with the *Go Racing* pack.

Pass the test, apply for your race licence, and

hey presto, you're a racing driver… And now the spending begins in earnest! That said, there's not much that can match the sheer thrill of being involved in a race, particularly the start and the first lap, and you'll soon be hooked. Don't forget your roots, though. Taking in a track day can be a great way to learn a circuit and even get some extra coaching when you're in the midst of a championship campaign. But just remember to leave your racing head at home…

ABOVE If you're the competitive type you might find you want to move up into the world of racing after you've taken part in a few track days. (*VW Racing*)

Appendix 1 Track day circuits

Sadly there's not nearly enough room to include all the track day venues here, so I've had to restrict this to the mainland UK *race* tracks – so apologies to Llandow, Bedford, Kirkistown and Jurby, all the airfield venues, and any others I've failed to mention; there just wasn't the room. But if you want to know about these, and many other venues all across Europe, check out the Circuit Guides referred to in Appendix 2.

Some will also notice that I've only included one layout of each track (again this is to do with space) but I've tried to include the one that will be more relevant to track day drivers.

As for the red segments of the track maps, they're the talking points. Every circuit has a talking point, whether it's the corner some say they take flat, or the place at which you're most likely to get into trouble. These, I hope, will help to give you a little of the flavour of each venue.

One of the best locations for a track day in the whole country has to be **Anglesey**. Right on the coast, and with views of Snowdonia, it really is just a bit special. And then there's the track, or rather the tracks. It was a brave move to replace a very popular layout with an international-standard circuit that incorporates four configurations. At the time of writing I've yet to visit, but by all accounts it's a move that has paid off big-time. Which means Anglesey should be as good a place for a track day as it ever was – which is very good indeed.
Where: Anglesey, LL63 5TF. Just off the A4080 near Aberffraw. Approach via the A55.
Contact: 01407 811400   (www.angleseycircuit.com)
Length: International GP (as shown) 2.1 miles; Coastal 1.55 miles; National 1.2 miles; Club 0.8 miles.
Talking point: On the International GP circuit shown, the left-hander approaching Rocket is said to be tricky, as it's blind over a crest, while on the shorter Coastal Circuit the link below the loop, known as the Corkscrew, is a great corner, I'm reliably informed. Stunning.

Brands Hatch is situated close to London and is a firm favourite with track day drivers. While track days on its awesome Grand Prix loop are rare and tend to be expensive, there are regular days on its shorter Indy circuit (as shown), which is just as undulating and quite challenging enough. Because it's short and twisty it can get a bit busy on a track day at Brands, but I wouldn't let that put you off as it's a wonderful track to drive. Circuit operator Motor Sport Vision will also sometimes run novice-only days here, which are ideal introductions to the world of track days.

Where: Kent, DA3 8NG. On the A20 about three miles from J3 of the M25.

Contact: 01474 872331 (www.motorsportvision.co.uk)

Length: Indy circuit (as shown) 1.2 miles; GP circuit 2.6.

Talking point: Has to be the awesome Paddock Hill Bend, which might very well be one of the best track day corners in the country. This fast right-hander has an apex that is over the brow and hence not visible to begin with, with the road then dropping away from you oh-so-very-steeply to a suspension-compressing exit. Paddock's not too difficult once you're used to it, but it's always massive fun. Classic.

BRANDS HATCH INDY

Another Motor Sport Vision run circuit, **Cadwell Park** is often referred to as the mini Nürburgring – which should tell you something about its unforgiving and challenging character. It's undulating, narrow, twisty, and utterly intoxicating. But you might want to make sure you've done a few track days at other venues before you pay this Lincolnshire circuit a visit, just to make sure you're at one with your car and you've an idea of how to tackle a corner. But visit it for at least one track day, you must, as it's a real gem.

Where: Lincolnshire, LN11 9SE. On the A153, 10 miles north of Horncastle, five miles south of Louth.

Contact: 01507 343248 (www.motorsportvision.co.uk)

Length: Full circuit (as shown) 2.1 miles; short circuit 1.5 miles.

Talking point: When a corner's called The Mountain you know you're in for a challenge. This starts off with a tricky double-apex left and then a right which demands precision – all topped off with a steep climb and brow, over which some cars will get airborne. Thrilling.

CADWELL PARK

Fast and challenging, **Castle Combe** has everything a driver could possibly want from a track day circuit, with technically challenging corners and high speed blasts. Indeed, this Wiltshire track was mega-quick up to the end of the '90s, but then some very big racing accidents led to moves to reduce the speed, which in turn led to the introduction of two chicanes. Happily, these have not really detracted too much from the high-speed nature of the circuit, and for the track day driver at least they might even add to the experience. Combe hosts a number of very good value 'Action Days' throughout the year, which are actually many track day drivers' introduction to the pursuit.

Where: Wiltshire, SN14 7EY. On the B4039, five miles west of Chippenham, and easily reached from J17 and J18 of the M4.

Contact: 01249 782417 (www.castlecombecircuit.co.uk)

Length: 1.85 miles.

Talking point: At race meetings the banking around Quarry Corner is packed with spectators, and that should tell you something. All the talk is over whether you brake before the brow at Avon Rise – which is the kink on the approach to Quarry – or after. It really depends on your car, but braking before the brow will make sense to begin with. But you should certainly resist the temptation to brake *on* the brow, where the car will go light. Tricky.

CASTLE COMBE

CROFT

With a mix of technical corners and sections that demand bravery if you're going to get through at the limit, **Croft** is a real driving challenge, and while it might be situated in the far north of England, track day drivers in the south should not be put off by the journey, as it really is worth paying this track a visit. There has been plenty of investment in the venue over the past few years, which means that facilities are pretty good, too.

Where: North Yorkshire, DL2 2PN. On the A167 just south of Darlington, not far from the A1(M).
Contact: 01325 721815 (www.croftcircuit.co.uk)
Length: 2.1 miles.
Talking point: Legend has it that the Jim Clark Esses were actually marked out by the great man himself, shaping the curves with a tractor and plough! If this is true, then the twice world champion did a fine job, as this is one of those 'flat or not' sequences that dominate paddock café chat at a track day. Enjoy!

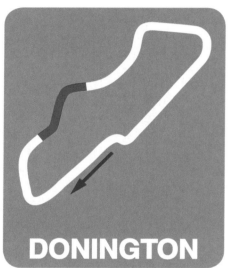

DONINGTON

If you think a bit of history adds to the occasion on a track day then you'll love **Donington Park**, for not only does it have an excellent grand prix car museum – well worth a visit during a track day lunch break – but it was also the venue for the famous Silver Arrow grands prix of 1937 and '38, and the site of what was perhaps Ayrton Senna's greatest F1 win in 1993. History aside, Donington is also an excellent track day venue, with the perfect mix of challenging corners, gradient, and run-off. The circuit's own Trakzone days are very popular and, unusually, will allow single-seaters.

Where: Leicestershire, DE74 2RP. Just off M1 (J23A or J24) and close to East Midlands Airport on the A453.
Contact: 01332 810048 (www.donington-park.co.uk)
Length: National (as shown) 2 miles; GP 2.5 miles.
Talking point: The Craner Curves running into the Old Hairpin seem to be everybody's favourite section of track at Donington. This is because you need to keep the car over to the left in the ultra-fast Craner to get the best line for the Old Hairpin. You really need to commit to Craner early, too, which catches some out, while it's flat for some cars and drivers, not quite for others. Exciting.

GOODWOOD

Goodwoood is all about nostalgia, its only race meeting of the year being the Revival. This takes place every September and draws crowds in the dress of the '40s, '50s and '60s, with Spitfires buzzing overhead and all sorts of cars from yesteryear out on track. But it's also used for a few track days throughout the year, so if you want to get to grips with a track that hasn't changed much in terms of character since the good old days, then this might be for you, although be warned that there can be tight restrictions on the number of cars allowed on track at one time.

Where: West Sussex, PO18 0PH. Close to Chichester. Approach via the A286 or A285.
Contact: 01243 755060 (www.goodwood.co.uk)
Length: 2.38 miles.
Talking point: At St Mary's a tricky and fast approach and negative camber at the apex make for a challenging sequence. It's also the place where Stirling Moss had his big crash in 1962. Historic.

One of the best places in the UK to watch Touring Cars, **Knockhill** is also a fantastic track day venue, with the circuit's own 'Speed Sundays' and novice track days proving popular. The Scottish track is up on a hill, which makes for wonderful elevation changes, but also some changeable weather. The circuit itself has an almost natural feel to it, which is obviously something to do with following the gradient, and although it's quite small there's plenty to keep you occupied right around the lap.
Where: Fife, KY12 9TF. From M90 take B914 then A823.
Contact: 01383 723337 (www.knockhill.com)
Length: 1.3 miles
Talking point: The SEAT curves are a long S bend where you need to get the exit just right. With a fast approach and late turn-in it can cause problems, while the gradient adds to the fun. Difficult.

Always in the shadow of Brands, yet **Lydden Hill** also has some things in common with that other Kent venue. Both are in a natural amphitheatre, and both have similar outlines featuring a looping hairpin. That's about the extent of the similarities, though, for where Brands is corporate and slick down to the last inch of painted kerb, Lydden is a little more, well… rustic. That's actually a good thing, though, for track days here tend to be relaxed affairs, while the circuit itself has enough of a mix of turns in its very short length to give you a real buzz.
Where: Kent, CT4 6RX. Just off the A2, a little west of Dover.
Contact: 01304 830 557 (www.lyddenracecircuit.co.uk)
Length: 1 mile.
Talking point: Something else Lydden shares with Brands, a Paddock Bend, though this time the quick right-hander is uphill to begin with and then crests a brow that makes the car go light – you might find your car will slide around quite a bit here. Marvellous.

Mallory Park is another short circuit, but also quite a deceptive one. To look at its outline you would think it was incredibly simple, yet there's enough variety in those four corners to tax the most experienced of track drivers. Curling around a couple of lakes, Mallory has a real parkland feel to it, while it also prides itself on being 'the friendly circuit'.
Where: Leicestershire, LE9 7QE. A47 from Leicester, or off J1 of M69, A5 then the A47.
Contact: 01455 842931 (www.mallorypark.co.uk)
Length: 1.35 miles.
Talking point: Gerards: a high-speed corner that seems to go on forever. It has a very long apex, which you can cling on to all the way round before letting the car sling shot out at the exit. Exhilarating.

OULTON PARK

Oulton Park is a proper challenge: high speed sections, a wide variety of corners, a long lap, and barriers that are close enough to the edge of the track to demand respect. Part of the Motor Sport Vision empire, Oulton was once the venue for top class international races such as the Gold Cup for Formula 1 cars, but these days its highest profile four-wheeled users tend to be the British Touring Car Championship guys, who always put on a good show when they get to Cheshire. It's also, without a shadow of a doubt, one of the best track day venues in the UK.
Where: Cheshire, CW6 9BW. On B5074, near Chester. Leave the M6 at J18 or J19.
Contact: 01829 760301 (www.motorsportvision.co.uk)
Length: International (as shown) 2.8 miles; Island 2.3 miles; Fosters 1.6 miles.
Talking point: Druids is a tough one to get just right, and you need to think hard about the entry. You can take a lot of speed through this right-hander, more than you might first think, but things are made difficult by a hidden exit. Challenging.

PEMBREY

One of my personal favourite circuits. **Pembrey**, which during the war was a Spitfire base, has everything a track needs to make it a challenge, except perhaps gradient. It's a technical circuit, with a good mix of fast, medium and slow corners – which makes it popular for testing, and up to the end of the '90s you could even find F1 teams testing here. Run-off's not bad, either, while there's not much that can beat Welsh countryside when it comes to track day backdrops. Oh, and it doesn't take as long to get to as you might think.
Where: Carmarthenshire, SA16 0HZ. Off the A484 close to Llanelli. Pembrey Circuit is ten miles from Junction 48 of the M4 Motorway.
Contact: 01554 891042 (www.barc.net)
Length: 1.5 miles.
Talking point: The corners really flow into each other at Pembrey, so it's difficult to separate them, but Paddock is at the heart of the twisting section of the track, a long left-hander which can be taken on a variety of lines, largely depending on what car you're in. It's further complicated by the need to get right back over to the left side of the track for the following bend. Technical.

ROCKINGHAM

Rockingham is different. Conceived as part of a bold plan to get UK race-goers hooked on American-style oval racing, its most obvious feature is the banked speedbowl, around which there's seating for 52,000 spectators. But there are also an astonishing 13 'road' circuit permutations, some of which are used for track days, and you can always be sure of a wide variety of turns and even a bit of gradient. Also, the pits and the facilities are very good at 'The Rock', all of which means it can make for a very good day's track driving.
Where: Northamptonshire, NN17 5AF. Near Corby, off the A6003. Leave M1 at J15 (from south) or J19 (from north).
Contact: 01536 500500 (www.rockingham.co.uk)
Length: International short (as shown) 2.5 miles.
Talking point: Much of the chat will be about the section that runs along the banking, particularly for first time visitors, but there's not much to driving it, other than thinking more about your momentum rather than your line. Different.

The venue for Britain's Formula 1 grand prix, **Silverstone** is a circuit everyone wants to drive at least once, particularly the longer layout used by the superstars, but the shorter options have plenty to offer, too, so don't count them out. Track days on the full circuit can be expensive anyway, but it's worth saving up for, as it's a wonderful track, and once you've driven it at speed in your own car you'll have a better understanding of the challenge that faces those who do it in the 200mph-plus projectiles that are F1 cars. As you would expect, facilities and pit garages are good, too.

Where: Northamptonshire, NN12 8TN. A43 near Towcester. Approach via J3 or J4 of the M40, or J2 of the M1.

Contact: 08704 588200 (www.silverstone.co.uk)

Length: Grand Prix (as shown) 3.2 miles; International 2.2 miles; National 1.6 miles.

Talking point: The Maggotts/Becketts/Chapel combination of corners is one of the best places in the world to watch a Formula 1 car, and it's also a hoot to drive through in your track day car. High speed and flowing, it's all about getting a fast exit on to the following straight, and it feels great when you get it just right. Legend.

With long straights, quick corners, and just a bit of the niggly slow stuff, **Snetterton** has it all. And, mainly because people seem to think it's in the middle of nowhere (it's actually very easy to reach from most parts of the UK), track days here can be good value. Another Motor Sport Vision circuit, Snet' has been upgraded over the past few years, and there are now plenty of gravel traps and lots of run-off – preferable to the grassy bank this writer hit *hard* in 1989. Ouch...

Where: Norfolk, NR16 2JU. Next to the A11, close to Thetford.

Contact: 01953 887303 (www.motorsportvision.co.uk)

Length: 1.9 miles.

Talking point: Coram is a wonderful corner, long and fast, and then right at the end you need to get it all settled for the tight chicane at Russell. Build up your speed slowly; this one can bite. Cool.

Thruxton is a very fast circuit which, sadly, doesn't seem to hold too many track days (although I've heard the 'experience' days are not bad). So if you do get the chance to go on a track day at Thruxton, then snap it up, because this is a great circuit. Many say it's difficult to learn, but that's probably because testing for racers is quite tightly restricted. What's for sure, though, is that it presents a mighty challenge, particularly with its high-speed sweepers 'out in the country' which give you a taste of how race tracks used to be back in the good old days.

Where: Hampshire, SP11 8PN. On the A303 near Andover. Take J8 of the M3.

Contact: 01264 882200 (www.barc.net)

Length: 2.4 miles.

Talking point: All the corners around the back are something else, and each leads into the other, but Goodwood is possibly the most challenging of them all, and it calls for confidence and commitment to take it flat. Quick.

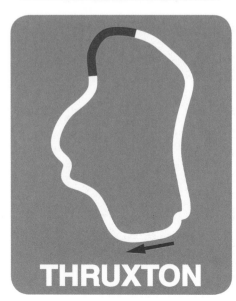

Appendix 2 Track day contacts

This is not a definitive list of everything that lives and breathes in the track day world (there just isn't the space for that), nor is it an endorsement of the companies and products mentioned. But, in most cases, they have been chosen on the basis of my personal experience, or from the experience of those in the track day scene I know and trust. I've limited each entry to a contact number (and website where applicable) and a brief comment where it might be helpful.

Although every effort was made to make sure this information was accurate at the time of writing, please bear in mind that circumstances, and contact details, will change over time.

TRACK DAY OPERATORS

Gold Track
01327 361 361   www.goldtrack.co.uk
email: info@goldtrack.co.uk
One of the better-known operators, and boss Calum Lockie is the chairman of the Association of Track Day Organisers, so you'll be in good hands.

Motor Sport Vision
0870 850 5014   www.motorsportvision.co.uk
The company that runs Brands Hatch, Oulton Park, Snetterton, Cadwell Park and Bedford Autodrome, and also some very good track days at these venues.

Javelin Track Days
01469 560576   www.javelintrackdays.co.uk
email: info@javelintrackdays.co.uk
A very popular outfit that puts on well-run days at both airfields and circuits.

Motorsport Events
0870 787 2116   www.motorsport-events.com
email: enquiries@motorsport-events.com
Runs lots of airfield days (many in the west country) plus the odd circuit day. Professional and friendly.

RMA
0845 260 4545   www.rmatrackdays.com
email: enquiries@rmatrackdays.com
A well-known operator that's been on the track day scene from the very beginning; organises well-run days in the UK and abroad, including the Nürburgring.

Trakzone
01332 819 503
www.donington-park.co.uk/trakzone
Donington's in-house operator runs regular and popular days, and evenings, at the circuit – one of the few places in the UK you can use a single-seater on a track day.

Circuit Days
01302 743827   www.circuit-days.co.uk
Runs good days, and also some imaginative road trips, which always seem to stop off at the 'Ring – quite right too!

Bookatrack
0870 744 1635   www.bookatrack.com
email: info@bookatrack.com
A well-known organisation which runs plenty of days throughout the year, including an annual visit to the awesome Spa circuit in Belgium.

Easytrack
07800 738413   www.easytrack.co.uk
email: team@easytrack.co.uk
Another well-known company with a reputation for quality track days and a wide selection of events.

Germantrack
07977 253065   www.germantrack.co.uk
A brand new organisation offering track days for German cars only – with plans to branch off into Japanese cars and 'retro' cars in the future.

Lotus on Track
07703 192 714   www.lotus-on-track.com
email: info@lotus-on-track.com
As the name suggests, this is a track day club that is, in its own words, 'run by Lotus enthusiasts for Lotus enthusiasts'. It's non-profit, too, so prices are quite low. If you've an Elise, Exige, 2-11, or any other Lotus, this could be just the thing.

Lydden Hill

01304 830557   www.lyddenracecircuit.co.uk
At the time of writing the small Kent venue was organising its own days. No passengers allowed, but great fun track days all the same, and a circuit that's definitely worth a visit.

Autotrack UK

01474 873942   www.autotrackdays.com
email: info@autotrackdays.com
Provider of track days on the Continent – including visits to the mighty Spa.

BHP Track Days

01342 837957   www.bhptrackdays.co.uk
Well-established company that runs a wide selection of days – including forays into France.

Knockhill

01383 723337   www.knockhill.com
The Scottish circuit runs its own good-value and innovative track days.

Castle Combe

01249 782417   www.castlecombecircuit.co.uk
'Action days' at the Wiltshire track are inexpensive and popular.

ORGANISATIONS
The Association of Track Day Organisers (ATDO)
0870 062 5352   www.atdo.co.uk

Motor Sports Association (MSA)
01753 765000   www.msauk.org

TRACK DAY ALTERNATIVES
Hill climb schools:
Gurston Down
01264 882215   www.gurstondown.org/school

Shelsley Walsh
01509 852253   www.shelsley-walsh.co.uk

RWYB:
Santa Pod
01234 782 828   www.santapod.co.uk

Drift days:
Javelin Track Days
(see 'Track Day Operators' above)

Motor Sport Vision
(see 'Track Day Operators' above)

Driftskool
www.driftskool.co.uk email: info@eurodrift.com

Drift Academy
www.driftacademy.co.uk
email: info@driftacademy.co.uk

The Nürburgring:
Ben Lovejoy's site
www.nurburgring.org.uk
Excellent website that tells you all you need to know about visiting the 'Ring.

Official site
www.nuerburgring.de

TRACK DAY INSURANCE
Moris.co.uk (Everritt Boles)
020 7709 9559
www.moris.co.uk

Competition Car Insurance
0115 941 5255
www.competition-car-insurance.co.uk

HIC
08451 291 291
www.hertsinsurance.com/trackdays

DRIVER COACHING
Instruction will be available at most track days, and that should be good enough if you're just starting out, but if later on you want to hire a coach for a day, or take part in a training course for track driving, you might want to try one of the following:

Mark Hales
www.markhales.com

Don Palmer
www.donpalmer.co.uk

Eugene O'Brien
01604 708766   www.eugeneobrien.co.uk

Sean Edwards
01491 572050   www.seanedwards.eu

Cadence Driver Development
01949 844 449   www.cadence.co.uk

Pentti Airikkala (left foot braking specialist)
01628 778808   www.leftfootbraking.com

TRACK GUIDES
UK Circuit Guide, European Circuit Guide
(including Nürburgring)
www.circuitguides.com

How to Drive... DVDs
www.markhales.com

TRACK DAY MEDIA
I've listed a few general magazines, and a website, here, but if you want to find a track day preparation expert for your car, then check out the adverts in the 'genre' mags (like *Japanese Performance*), or the one-make mags (such as *VW Golf+*), which should point you towards marque specialists.

www.pistonheads.com

evo magazine
www.evo.co.uk

Practical Performance Car magazine
www.ppcmag.co.uk

Performance Tuner magazine
www.performancetuner.co.uk

TRACK DAY CAR HIRE
Many of the bigger operators will have track cars they will hire out, so check out their websites, and it also might be worth trying these:

Track-Club
01480 477000   www.track-club.com

U Drive Cars
www.udrivecars.com/track_day_car_hire

RETAILERS
For racewear and go-faster goodies there are plenty to choose from, but the two below are among the best known:

Demon Tweeks
01978 664466   www.demon-tweeks.co.uk

GPR
0844 880 1750   www.gprdirect.com

OTHER
www.racecarsdirect.com

CG Lock
0161 832 3786   www.cg-lock.co.uk

FURTHER READING
Competition Car Suspension – a practical handbook, Alan Staniforth (Haynes)
Very useful if you want to understand more about suspension.

Race and Rally Car Source Book, Alan Staniforth (Haynes)
A great way to start to get your head around sorting a car for the track.

Four-Stroke Performance Tuning, A. Graham Bell (Haynes)
For when you get to the stage that you want a few more horses under the bonnet.

Speed Secrets, Ross Bentley (Motorbooks)
This cracking book has now developed into a series of six volumes, each of which examines a different facet of the racing driver's art.

Going Faster!, Carl Lopez (Robert Bentley)
An extremely useful tome on track driving from the team at the Skip Barber Racing School.

Ayrton Senna's Principles of Race Driving (Ayrton Senna)
Not big on detail, but interesting all the same, and quite wide ranging.

Bob Bondurant on High Performance Driving (Bob Bondurant and John Blakemore)
I've not seen a copy of this for a while but it's a book that really helped me when I was racing.

Index

Index